BACTERIOLOGY RESEARCH DEVELOPMENTS

COLIFORMS

OCCURRENCE, DETECTION METHODS AND ENVIRONMENTAL IMPACT

Bacteriology Research Developments

Additional books in this series can be found on Nova's website under the Series tab.

Additional e-books in this series can be found on Nova's website under the e-book tab.

BACTERIOLOGY RESEARCH DEVELOPMENTS

COLIFORMS

OCCURRENCE, DETECTION METHODS AND ENVIRONMENTAL IMPACT

GEOFFREY MCCOY
EDITOR

nova publishers
New York

Library of Congress Cataloging-in-Publication Data

ISBN: 978-1-63483-542-8

Library of Congress Control Number: 2015946957

Published by Nova Science Publishers, Inc. † New York

CONTENTS

Preface **vii**

Chapter 1 Impact of Biofilms in Drinking Water Contamination by *Escherichia coli* Coliform **1**
Anabela Borges and Manuel Simões

Chapter 2 Methods of Analysis for Bacterial Contamination in Environmental Waters **27**
Robert G. Price and Dirk Wildeboer

Chapter 3 Testing of Coliformic Bacteria for Water Departments Using Automated Sampling and Analysis **53**
Elias Hakalehto, Anneli Heitto, Osmo Hyyrynen, Juha Jokelainen, Ari Jääskeläinen and Lauri Heitto

Chapter 4 Prevalence of *Escherichia col*i and Outbreaks from Ready to Eat Fresh Fruits and Vegetables **71**
Agni Hadjilouka, Spiros Paramithiotis and Eleftherios H. Drosinos

Chapter 5 Sources and Survival of Coliform Bacteria in Temperate Freshwaters: A Review **107**
E. Pip

Index **125**

PREFACE

Coliforms are widely used as indicators of water quality and sanitary status, and levels are regulated as standards and legislation in many jurisdictions. The group of organisms known as coliform bacteria are commonly used to gauge quality and potential for health risk of potable and recreational waters, and are stimulated in water quality standards and legislation of many jurisdictions worldwide. The contents of this book includes an overview about the involvement of biofilms in waterborne diseases; currently employed methods of analysis for coliform bacteria; the testing of coliformic bacteria for water departments using automated sampling and analysis; the prevalence of E.coli outbreaks linked to the consumption of certain products; and finally, a review of the sources of coliform bacteria in freshwaters.

Chapter 1 – Waterborne diseases continue to be an important cause of illness and death with particular incidence in developing countries. Indeed, the use of contaminated drinking water (DW) has been linked with several health problems worldwide. So, the main goal of water companies is to ensure that water is microbiologically and chemically safe for consumers. Many of the problems in DW distribution systems (DWDS) are microbiological and thus the prevention of water associated problems has mainly focused on these contaminations. There are several public health relevant microorganisms that may be transmitted through contaminated DW and infect humans. Even more worrying is the fact that waterborne bacterial pathogens have the potential to become components of sessile microbial communities. Biofilms can act as environmental reservoirs of coliforms (intestinal and non-intestinal origin) and other clinically relevant pathogens, representing a potential source of water contamination. The presence of biofilms in DWDS constitutes special concern because they provide the ideal environment for pathogens to survive and persist for long time in distribution networks. As biofilms are recalcitrant to treatment, they also contribute to the reduced efficiency of disinfection programs. This chapter provides an overview about the involvement of biofilms in waterborne diseases. In addition, the impact of the presence of biofilms in DWDS and on the organoleptic quality of the distributed water will be also discussed. As *Escherichia coli* is an indicator of fecal contamination and some of their variants, enterotoxigenic and enterohemorrhagic, are one of the major causes of water-related outbreaks, their role will be also highlighted.

Chapter 2 – Water pollution by faecal contamination is a serious problem due to the potential for contracting disease if the water is consumed untreated. Coliform bacteria *Escherichia coli* and *Enterococci* are used as primary indicators of contamination in fresh and marine water. The presence and levels of these organisms are indicative of the presence of a number of pathogenic organisms of animal origin and thus for a potential threat to human health. The acceptable levels of indicator organisms are defined by legislation and vary for drinking, river, well and marine water. This review will consider currently employed methods of analysis for coliform bacteria, compare them with recently developed procedures and evaluate advantages and limitations of different methods. Emphasis will be placed on methods which can be used on-site with laboratory procedures as confirmatory procedures. Established culture methods use chromogenic substrates for β-D-glucuronidase to identify *E. coli* and β-D-galactosidase to detect *Enterococci*. Rapid non-culture based fluorimetric procedures have been developed that can be used on-site. In addition to culture and biochemical procedures, advances have been made in developing genomic, proteomic and metabolomic methods. Methods must be quantitative in order to comply with legal norms and in this regard the potential involvement of biosensor technology will be of great value in successfully transferring laboratory procedures to the field. The Chapter will conclude with a recommendation of the most valuable laboratory and on-site methods for water analysis given the constraints of the current legal framework governing acceptable water quality.

Chapter 3 – In municipal or household water production and in its monitoring, there is a need to protect the consumers from dangerous microbial impacts. It is important to find out even the smallest concentrations of hazardous organisms or their toxins. In order to screen huge masses of water flow, enrichment technologies and indicator organisms are used according to international guidelines. Also, automated sampling and analysis—as well as data transfer—are warranted. The PMEU technologies have been developed for enhanced microbial cultivation and detection. The Coliline PMEU version is developed for automated water monitoring.

Chapter 4 – Food poisoning outbreaks linked to disease-causing strains of the bacterium *Escherichia coli* are usually associated to meat products. However, 20-30% of *E. coli* infections are caused by consumption of fresh produce due to mishandling at processing facilities, transportation and food service establishments. *E. coli* has been isolated from a variety of products and its prevalence has been reported to range from 2.5% to 93.3% in leafy greens, spinach, lettuce, cucumbers, rocket, coriander, celery, etc. At the same time various fresh fruits and vegetables have been implicated in outbreaks of *E. coli* infections, with the 1996 Japan outbreak linked to consumption of white (daikon) radish sprouts being the world's largest reported outbreak and the 2011 E.U. outbreak linked to seed sprouts being the outbreak with the highest mortality rate. In this chapter prevalence of *E. coli* in unprocessed and minimally processed fruits and vegetables as well as *E. coli* outbreaks linked to the consumption of such products are addressed.

Chapter 5 – The sources and survival of coliform bacteria in freshwater are reviewed. Coliforms are widely used as indicators of water quality and sanitary status, and levels are regulated as standards and legislation in many jurisdictions. However both total (TC) and fecal (FC) coliforms may originate from various environmental sources besides human intestinal flora, and many are nonpathogenic. In surface water, TC and FC may survive for extended periods depending on physical and chemical factors such as ultraviolet light, temperature, pH, turbidity and organic matter. Biotic factors such as protozoa, bacteriophages and algae may also affect survival.

In: Coliforms
Editor: Geoffrey McCoy
ISBN: 978-1-63483-542-8

Chapter 1

IMPACT OF BIOFILMS IN DRINKING WATER CONTAMINATION BY *ESCHERICHIA COLI* COLIFORM

Anabela Borges[1,2,3] and Manuel Simões[1]
[1]LEPABE-Department of Chemical Engineering,
Faculty of Engineering, University of Porto, Porto, Portugal
[2]CIQUP-Department of Chemical and Biochemical,
Faculty of Sciences, University of Porto, Porto, Portugal
[3]CECAV-Veterinary and Animal Science Research Center,
Veterinary Science Department,
University of Trás-os-Montes e Alto Douro,
Vila Real, Portugal

ABSTRACT

Waterborne diseases continue to be an important cause of illness and death with particular incidence in developing countries. Indeed, the use of contaminated drinking water (DW) has been linked with several health problems worldwide. So, the main goal of water companies is to ensure that water is microbiologically and chemically safe for consumers. Many of the problems in DW distribution systems (DWDS) are microbiological and thus the prevention of water associated problems has mainly focused on these contaminations. There are several public health relevant microorganisms that may be transmitted through contaminated DW and infect humans. Even more worrying is the fact that waterborne bacterial pathogens have the potential to become components of sessile microbial communities. Biofilms can act as environmental reservoirs of coliforms (intestinal and non-intestinal origin) and other clinically relevant pathogens, representing a potential source of water contamination. The presence of biofilms in DWDS constitutes special concern because they provide the ideal environment for pathogens to survive and persist for long time in distribution networks. As biofilms are recalcitrant to treatment, they also contribute to the reduced efficiency of disinfection programs. This chapter provides an overview about the involvement of biofilms in waterborne diseases. In addition, the impact of the presence of biofilms in DWDS and on the organoleptic quality of the distributed water will be also discussed. As *Escherichia coli* is an indicator of fecal contamination and some of their variants, enterotoxigenic and enterohemorrhagic, are one of the major causes of water-related outbreaks, their role will be also highlighted.

Keywords: biofilms, coliforms, drinking water distribution systems, *Escherichia coli*, waterborne diseases

1. Introduction

Water is undoubtedly essential for life and one of the most common and important chemical compounds on Earth. Of the global water quantity, only 1.4×10^9 km^3 (~2.6%) is freshwater. The availability of safe drinking water (DW) is definitely an indispensable requisite for survival and well-being [1]. So, the provision of harmless DW is a basic human right that is considered a top priority of the communities and regulatory agencies [2]. However, today there are yet many people in some Earth areas that do not have access to proper DW. In addition to water quality, the scarcity is another important issue with which they are continually faced. It is estimated that globally more than 1 billion of people have no access to sufficient safe DW [3].

With the progressive increase in urbanization and population growth the necessity to implement public water supply networks assumed great importance [4]. In the highly populated areas, the amount of waste, wastewater and other types of contamination increases, compromising the access to fresh and DW. The rapid expansion of industry, wherein the chemical industry deserves special attention, has also constantly resulted in contamination of the different types of natural water sources [1]. Besides, the inadequate planning, design and construction of the DW distribution systems (DWDS) associated to an evident lack of maintenance and control, leads frequently to water quality deterioration [5]. In this sense, the public water supply can be a risk to the health of a large number of people if contamination occur [4]. Indeed, water is considered one of the major reservoirs and/or vehicle of disease transmission. All the aspects mentioned above, and therefore the utilization of insalubrious DW has been linked with a wide range of health problems [6]. These water-related illnesses can be acquired by any person that contact with inappropriate DW, but its incidence is more pronounced on immunocompromised persons (e.g., AIDS patients, patients receiving chemotherapy or organ transplantation) and in those that are more susceptible like children's and elderly [7]. The World Health Organization (WHO) highlights that the quality of DW is a powerful environmental determinant for health. So, the management of its safety was undoubtedly a central focus for centuries and remains the base to ensure the prevention and control of water-associated health problems [5]. In order to maintain a more rigorous control of water quality, the WHO periodically publishes international standards for DW [2].

The main goal of this chapter is to provide an overview about bacterial biofilm development in DWDS and its impact on the deteriorations of water quality. Besides, their potential risk to human health will be also highlighted. The role that biofilms play in the persistent occurrence of the *Escherichia coli* coliform, including pathogenic and non-pathogenic forms, will be also explored. Finally, new measures for the control of biofilm formation will be briefly presented.

2. Microbiological Contamination of Drinking Water

The main goal of water authorities is to ensure that the water that reaches the consumer is safe for consumption and free from any substances that may be dangerous to health [8]. Water contamination can have chemical or microbiological origin, but the majority of the problems in DWDS have microbial nature [9].

Microorganisms can enter in the distribution systems through a wide range of events, such as treatment processes or lacunas of the distribution systems infrastructure [7]. With respect to microbiological contamination, waterborne diseases are considered one of the most important water related problems that present a serious global threat. They are usually generated by the use of DW contaminated with human or animal feces harboring opportunistic pathogens, like bacteria, protozoa, virus and helminths [2]. These waterborne agents may display different characteristics, behavior and resistance to antimicrobials. Among the referred microorganisms, bacteria are normally the group of pathogens with higher sensitivity to disinfection. However, it is known that some pathogenic bacteria exhibit elevated resistance to chlorine, the most typical product used for DW disinfection. Therefore, if these bacteria are not totally removed by disinfection processes and reach the consumers tap, they can cause disease outbreaks [10-12]. Indeed, the proportion of waterborne disease outbreaks associated with distribution system failures has been increasing over the years [13].

The occurrence of outbreaks affects not only the developing countries, but also those that are more developed. Whereas lack of safe drinking water supply in developing countries is mostly because of economic constraints, in developed countries health risks by waterborne pathogens may occur through mismanagement of freshwater resources, technological failure and/or inappropriate detection measures [14]. Although in developed countries water-related diseases are rarely deadly, several authors report that DW consumption containing pathogenic bacteria has numerous health risks [12, 15]. There are a number of waterborne bacterial pathogens that can be transmitted by water, including *Salmonella*, *Campylobacter*, *Vibrio*, *E. coli*, *Shigella, Helicobacter*, *Legionella*, *Pseudomonas* and *Aeromonas* [1, 16, 17]. In the last years, waterborne outbreaks caused by *E. coli* O157:H7, *Campylobacter* spp., *Helicobacter pylori*, *Pseudomonas aeruginosa*, *and Salmonella paratyphi* have been reported in several countries [3, 18]. Almost all enteric and opportunistic pathogens that are transmissible by the fecal-oral route can be also transmitted through water. However, the critical characteristics of an organisms that define the risk to occurs waterborne outbreak of disease are the rate of inactivation during disinfection and the infective dose [19]. According to a statement of the WHO, between 2000 and 2007 in fourteen European countries, occurred 354 outbreaks of waterborne diseases related to contaminated DW that resulted in over 47 617 episodes of illness [20].

Gastroenteritis is the most well-known disease associated with waterborne outbreaks in developed countries, however, there are many other complications that can occur (cholera, typhoid fever, meningitis, encephalitis, dysentery, hepatitis, legionellosis, pulmonary illness, poliomyelitis, leptospirosis, giardiasis and salmonellosis [2, 21]. For instance, it is estimated that diarrheal disease accounts for 4.1% of the total global burden of disease and kill 1.8 million of people every year [5].

Chlorination is a universal method applied in many countries to DW disinfection, however water-mediated disease outbreaks continue to happen worldwide [22]. Effectively, even when the water treatment is of high quality, their characteristics can be seriously deteriorated during storage and distribution, if precautions are not taken [23]. As the last barriers prior to DW delivery to the consumer's tap are the distribution systems, episodes of contamination post-treatment are critical and tend to attract considerable attention [4]. Therefore, the development of robust DWDS and the use of adequate disinfection measures are very important to reduce the incidence of waterborne diseases. However, despite primary and secondary disinfection steps, several coliform and opportunistic pathogenic bacteria are occasionally found within water systems [24]. This occurrence is mainly due to the growth of microorganisms within sessile communities named biofilms, which develop on the inner surfaces of distribution networks [1, 25]. They are structured microbial communities surface-attached by a self-produced matrix of extracellular polymeric substances (EPS) [26-29]. Biofilms are common in water distribution systems being recognized as a normal part of them. As biofilms facilitate the survival of potentially pathogenic microorganisms, and consequently cause a variety of problems in water quality, their presence in DWDS constitutes a concern for the public health community [3].

3. Biofilms and Problems Associated with Their Formation in Drinking Water Distribution Systems

From the microbiological point of view, biocorrosion, nitrification, occurrence and persistence of pathogenic organisms and also biofilm formation, are the main problems reported in DWDS [30-33]. Among them, biofilms have been intensively studied in the last decades, due to their recognized impact in the health and water industry, where they assume considerable hygienic, operational and economical relevance [3].

Biofilm formation is suspected to be the main factor responsible by the presence of microorganisms in DWDS, constituting a reservoir for pathogens even after treatment cycles [3]. Besides, the presence of biofilms may lead to a number of undesirable effects on water organoleptic properties, altering the taste, odour and colour [34-36]. In the same way, they are responsible for the increased deterioration and obstruction of metallic pipes, which lead to the decrease in water velocity and thus hinder an efficient cleaning and/or disinfection of the water distribution systems [3]. Hence, when detachment of some portions of the biofilm occurs, which in turn enter in bulk flow, they are responsible for various waterborne diseases outbreaks [37]. In fact, biofilm clumps can be an efficient means to deliver high infective dose of diverse pathogens, when are taken up by host [38]. Another concern related with biofilm dispersion is the rapid colonization of clean areas that in some way contributes to the persistence of bacterial population in the system [3].

The conventional approach to control microbial growth in DWDS, i.e., the maintenance of residual disinfectant, is often ineffective to eliminate biofilm associated bacteria as demonstrated in diverse works [9, 39]. Studies performed in real distribution systems have shown that the maintenance of residual levels of chlorine were insufficient to control biofilms in DWDS [40]. In the same way laboratory-based studies demonstrated that bacteria attached to surfaces were 150 to over 3000 times more resistant to chlorine than the same cells in suspension [41]. Keevil et al., [42] reported that coliform bacteria in biofilm survived to 12 ppm of free chlorine exposure during long time.

It is well known that the growth of the microorganisms as sessile communities can confer them numerous advantages compared to planktonic or non-biofilm associated cells. These include higher protection against environmental stress, predators and chemicals (e.g., disinfectants and antibiotics); increased access to nutrients; closer proximity between cells that facilitate the mutualistic/synergistic associations; and also gene transfer that permit the acquisition of antibiotic resistance genes [43-46]. Microbial community diversity and interspecies relationship within biofilms are also two important factors that can influence the resistance of pathogens to disinfection [9]. Indeed, multispecies biofilms have demonstrated more resistance to biocides and disinfectants than single-species biofilms [24, 47-49].

The mechanisms behind the observed resistance of biofilm cells to disinfection are not complete understood, although some hypotheses have been attributed. One of the described resistance mechanisms is the disinfectant interaction (e.g., chlorine) with biofilm constituents, including microbial cells and EPS, which result in it consumption and also limit their penetration into the biofilm [39, 50]. Additionally to chlorine decay, other important characteristics that have been described to explain why bacteria in biofilm are more protected comprise: altered bacterial metabolic state (i.e., reduced growth rate) in part due to nutrient/oxygen limitation; presence of persister cells; expression of specific biofilm resistance genes; and biofilm specific phenotype [51-53].

Various investigations have demonstrated that the prevalence of pathogens in distribution systems that harbor biofilms can reach surprising levels [54-56]. Although wrongly, the routine microbiological monitoring only includes the search of planktonic indicator organisms, and does not take into account the presence of biofilms in the distribution system, where it is estimated that up to 95% of the bacterial biomass may be found [57]. As microbial growth in biofilms can affect directly the cell density, in the bulk phase, biofilms are greatly relevant for water quality as they contribute to their deterioration. Although, the elimination of these microbial consortiums from DWDS is almost impossible some precautions can be taken in order to prevent and control their growth. They include the control of the nutrient content in water by minimizing the entrance of organic matter in the distribution systems; the concentration of residual disinfectant throughout the distribution system; the hydrodynamic conditions of the network in order to prevent the water stagnation and sediment accumulation within distribution systems; the material that is used in the construction of the pipes; the environmental factors (e.g., pH and temperature) and the diversity of microorganisms [3]. Figure 1, provides an overview of the main factors related with biofilm formation in DWDS and some measures suggested for their control.

4. *Escherichia coli* as an Indicator of Microbial Water Contamination

The search of fecal indicator microorganisms was implemented since 1990`s, because it is not feasible to investigate the presence of all known waterborne pathogens and therefore assess whether water is microbial safe for drinking [2]. *E. coli* is a normal inhabitant of the intestine of man and other warm-blooded animals and is excreted in the feces. For this reason, its detection in the environment is indicative of pollution, being an important marker for water hygiene (except in the tropical environments) [58, 59]. So, the microbiological parameters for routine monitoring of drinking water recommend *E. coli* and thermotolerant coliforms as indicators of the effectiveness of disinfection processes, and as index organisms for the potential presence of fecal contamination [2, 15, 59].

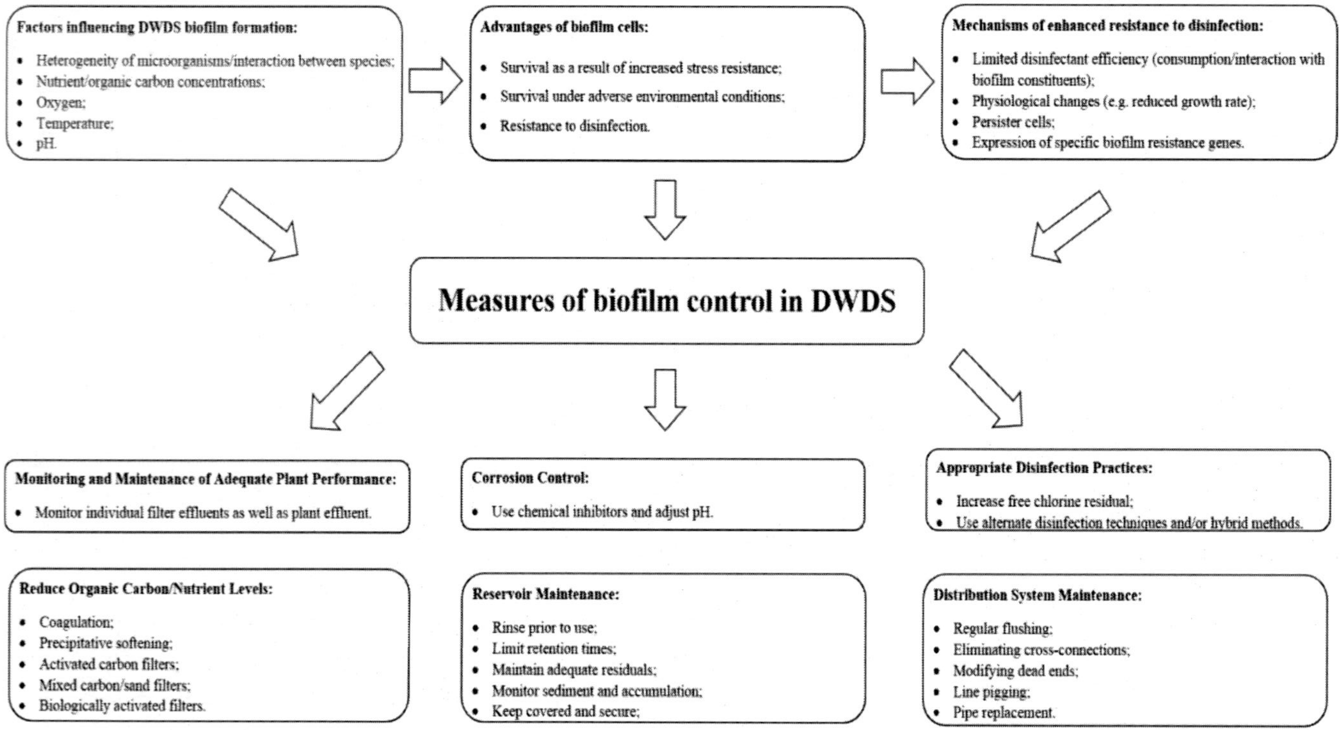

Figure 1. Factors that influence biofilm formation and growth of the biofilm cells in DWDS, disinfection resistance mechanism and some precautions that can be taken in order to prevent and control biofilms.

The presence of *E. coli* and other coliform bacteria in DWDS have importance not only as potential pathogens but also as indicators of the possible presence of other hygienically relevant microbial pathogens [24].

4.1. *Escherichia coli* - Enterovirulent Group

In general, the *E. coli* strains that colonize the humans are harmless commensals. However, within the species there are strains highly pathogenic, due to the occurrence of virulence factors such as enteroadhesins or enterotoxins, which cause distinct syndromes of diarrheal disease [59]. Indeed, pathogenic *E. coli* are described as one of the major cause of diarrheal disease worldwide, occasionally with fatal outcomes [19, 60]. The main categories include, enteropathogenic (causes a pedestal-like structure and produces attaching-and-effacing (A/E) histopathology), enterotoxigenic (produces enterotoxins), enteroinvasive (characterized by the invasion of the colonic epithelium), enterohemorrhagic (causes severe abdominal cramps and bloody diarrhea), enteroadherent (exhibit fimbrial mediated adherence to Hep-2 cells) and enteroaggregative (distinguished by prominent auto-agglutination of bacterial cells) *E. coli* [59-61]. Some of these *E. coli* strains, although with less frequency, can be found within potable water systems, constituting a public health risk, since there is no standards for the search of their presence [60].

In developed countries like United States the most described waterborne outbreaks are related with *E. coli* strains belonging to the enteropathogenic and enterotoxigenic classes [59, 62, 63]. Other outbreaks of enterotoxigenic *E. coli* were reported in Japan [64].

In addition to diarrheal disease, pathogenic *E. coli* strains are also associated with a variety of diseases. Each specific type of the pathogenic categories mentioned earlier, causes disease through different mechanisms and each disease presents different clinical symptoms. The diverse *E. coli* serotypes, the interaction with host cells, associated illnesses and symptoms are presented in Table 1.

The most problematic and pathogenic *E. coli* strain is enterohemorrhagic *E. coli* O157:H7 and therefore it is also the best studied serotype. However, there are more than 100 other serotypes of *E. coli* [1, 65]. *E. coli* O157:H7 is known for their high capacity to cause severe bloody diarrhea (hemorrhagic colitis) and renal failure (hemolytic uremic syndrome) in humans [66]. Waterborne outbreaks with large extension caused by enterohemorrhagic *E.*

coli O157:H7 were documented [18, 67, 68]. For example, in 49 states of the USA was reported 73 000 cases of illness caused by *E. coli* O157:H7 [69].

The knowledge of the pathogens distribution and survival strategies within DWDS, and the application of methodologies that can detect not only the presence, but also the viability and infectivity of the pathogens permits to assess with more precision the waterborne risks [19]. Indeed, some of the bacterial strains in DWDS are viable but non-cultivable, and most of them are attached to the inner surfaces of the pipes.

In a recent survey performed by Oliver [70], all known waterborne bacterial pathogens can enter in a state of non-cultivability without loss of viability. Also, environmental biofilms are suspected to harbor substantial numbers of non-cultivable cells that can survive over extended periods of time [71].

The quantification or even the isolation of *E. coli* O157:H7 has rarely been achieved by traditional detection methods [61]. Due to the clinical significance that enterohemorrhagic *E. coli* have assumed in the medical sector and the various waterborne outbreaks documented, a diverse range of detection methods (culture based and non-culture based) for DW samples have been developed [61, 72]. They include: differential and enriched media; CTC-fluorescent antibody; immunoassay enzyme-labelled; phages; immunomagnetic separation; real time PCR; microarrays analysis; multiple-locus variable-number tandem repeat analysis and pulsed-field gel electrophoresis [60, 61].

4.2. *Escherichia coli* Occurrence in Drinking Water Distribution Systems Linked to Biofilms

The presence of biofilms in distribution systems is one of the sources attributed to the recurrent occurrence of coliforms and other pathogenic bacteria in DW [36, 40]. Diverse studies evidenced that coliform bacteria, including *E. coli*, associated to biofilms are able to survive and multiply even under low nutrient/temperature conditions of the distribution systems [73-75]. Among these are the resurgence of total coliforms from distribution systems in which water coming out of the treatment unit was previously negative for coliforms [74]. Another study demonstrated that *E. coli* was recovered from DW distribution pilot plants, although the calculated inoculum should have theoretically washed out [75].

Table 1. ***E. coli*** **enterovirulent groups, serotypes, host cell interaction, associated illnesses (time to onset/duration) and symptoms. Adapted from Percival et al., [58, 60]**

Eenterovirulent group	Serogroup	Host cell interaction	Illness	Time to onset of illness	Duration of illness	Symptoms
Enteropathogenic (EPEC)	055 H6, NM 086 H34, NM 0111 H2, H12, NM 0119 H6, NM 0125ac H21 0126 H27, NM 0128 H2, H12 0142 H6	EPEC attach to intestinal mucosal cells causing cell structure alterations (attaching and effacing) EPEC cells invade the mucosal cells	Enteritis in infants Traveller's diarrhea	17 to 72 h (average 36 h)	6 h to 3 days (average 24 h)	Severe diarrhea in infants which in many persist for more than 14 days. Also fever, vomiting and abdominal pain. In adults, severe watery diarrhoeal, with prominent amounts of mucus without blood, and nausea, vomiting, abdominal cramps, headache, fever and chills.
Enterotoxigenic (ETEC)	06 H16, 08 H9, 011 H27, 015 H11, 020 NM, 025 H42, NM, 027 H7, 078 H11, H12 0128, H7 0148, H28, 0149 H10, 0159 H20, 0173 NM	ETEC adhere to the small intestinal mucosa and produce toxins that act on the mucosal cells	Diarrhea, vomiting and fever Traveller's diarrhea	8 to 44 h (average 26 h)	3 to 19 days	Watery diarrhea, low grade fever, abdominal cramps, malaise, nausea. When severe, causes cholera-like extreme diarrhea with rice waterlike stools, leading to dehydration.

Table 1. (Continued)

Eenterovirulent group	Serogroup	Host cell interaction	Illness	Time to onset of illness	Duration of illness	Symptoms
Enterohemorrhagic, (EHEC)	026 H11, H32, NM 055 H7 0111ab H8, NM 0113 H21 0117 H14 0157 H7	EHEC attach to and efface mucosal cells and produce toxin	Shigella-like dysentery (stool contain blood and mucus) Hemolytic uraemic syndrome	3 to 9 days (average 4 days)	2 to 9 days (average 4 days)	Hemorrhagic colitis (HC): sudden onset of severe cramps, abdominal pain, grossly bloody diarrhea, vomiting, no fever. Hemolytic uraemic syndrome (HUS): prodrome of bloody diarrhea, acute renal failure in children, thrombocytopenia, acute nephropathy, seizures, coma, death. Thrombotic thrombocytopenic purpura (TTP): similar to HUS but also fever, central nervous system disorders, abdominal pain, gastrointestinal hemorrhage, blood clots in the brain, death.

Eenterovirulent group	Serogroup	Host cell interaction	Illness	Time to onset of illness	Duration of illness	Symptoms
Enteroinvasive (EIEC)	028ab NM 029 NM 0112ac NM 0124 H30, NM 0136 NM 0143 NM 0144 NM 0152 NM 0159 H2, NM 0164 NM 0167 H4, H5, NM	EIEC invade cells in the colon and spread laterally, cell to cell	Shigella-like dysentery	8 to 24h (average 11 h)	Days to weeks	Profuse diarrhea or dysentery, chills, fever, headache, muscular pain, abdominal cramps.
Enteroaggregative (EAEC)	03 H2 015 H18 044 H18 086 NM 077 H18 0111 H21 0127 H2 0? H10	EAEC bind in clumps (aggregates) to cells of the small intestine and produce toxins	Persistent diarrhoeal in children	-	-	Watery mucoid diarrhea, without blood and afebrile.
Eenteroadherent (EADEC)	Not yet established	Fimbrial and non-fimbrial adhesions identified	Childhood diarrhea	-	-	Age-dependent diarrhea, fever, vomiting, abdominal pain and dehydration.

In a work performed by Williams and Braun-Howland [76], injured and noninjured *E. coli* strains and other coliforms within biofilms in a model distribution system, presented resistance to treatment with recommended levels of disinfectants (hypochlorous acid and monochloramine). Besides, these authors showed that *E. coli* within biofilms not only survived but also maintained their metabolic activity. In this line, it is possible to assume that the release of coliform bacteria from biofilms can be one of the causes of sporadic or intermittent events of coliform occurrence during routine DW analysis. Consequently, it is necessary to have additional precautions, because the presence of biofilms in DWDS can "camouflage" the occurrence of true system failures, delaying the take of adequate microbial growth control measures. Besides, the formation of biofilms in distribution systems reduce the accuracy of the indicator functions of coliforms [39, 76].

No evidence exists about the higher capacity of enterovirulents *E. coli* to resist to water treatments comparatively to non-pathogenic *E. coli*. However, concerns exist about the protective role that biofilms can have to these strains [58, 60]. This occurrence is particularly worrying in the case of enterohemorrhagic *E. coli* once it requires low infectious doses (fewer than 100 organisms) to trigger a disease, and thus the potential risk of infection from biofilms in water is significant. It is evident that the persistence of *E. coli* in natural environments is favored through its ability to grow as a biofilm in the presence of other microbial species that lead to a potentially cooperative coexistence [77]. For instance, Bauman et al., [78] demonstrated that *E. coli* O157:H7 adhered more effectively to reactors conditioned with a drinking water biofilm. In a study performed by Moreira et al., [79], *E. coli* strains isolated from freshwater lakes demonstrated higher capacity to form biofilms than strains isolated from humans. The high biofilm-forming capacity of the environmental strains can be explained by the advantages that they can obtain when grown as sessile communities. Recent studies revealed the presence of enterotoxigenic *E. coli* biofilms in household water in Dhaka-Bangladesh [80]. Besides, high numbers of *E. coli* in biofilms have also been found in a drinking water treatment plant in Romania [81].

5. Alternative Disinfection Techniques to Biofilm Control in Drinking Water Distribution Systems

In the last century, the microbiological safety of public water supplies has experimented improvements, particularly in developed countries, due to significant advances in water treatment plans. However, incidents involving water supplies have been documented in recent years, in countries with the most advanced water treatment systems [4, 15, 18]. These gaps are mainly due to the incapacity to maintain an adequate level of disinfectant throughout the DWDS. The maintenance of a residual amount of disinfectant (known as secondary disinfection) is designed to prevent the re-growth of bacteria, eliminate any pathogens that can enter into the system, and prevent the growth of biofilms within that system [82]. Chlorine is still the most commonly used disinfectant in DWDS. According to the WHO guidelines for drinking water quality a residual level of 0.2 mg/L of chlorine need be maintained along of the distribution systems, with a maximum of 5.0 mg/L [1, 2].

As referred above, the level of disinfectant generally employed in DWDS are insufficient to prevent the growth and development of microbial biofilms and once biofilms are established on pipe surfaces, their elimination is almost impossible [83]. In this context, there is an urgent need to develop new effective disinfection techniques and alternative methods in order to diminish biofilm accumulation in DWDS. Besides, it is also of great importance to minimize the environmental (e.g., generation of residual chemical pollutants) and public health (e.g., high chlorine concentrations increase the production of harmful by-products) impacts of the traditional practices of disinfection [2, 3]. In fact, the use of disinfectants, like chloride, chlorine dioxide and chloramines, is not the only strategy that can be applied to preserve water safe until it reaches the consumers. For instance, membrane technology is used in some countries to treat DW. Occasionally, they also use a mechanical strategy named pipe flushing that allows to remove the sludge and sediments from pipe walls, eliminates tastes and odors, reduces turbidity and restores the disinfectant residual concentration, affording a better control of microbial regrowth [84, 85]. Moreover, the use of chlorine can be significantly reduced if combined with other strategies [3].

Other non-chemical alternatives for microbial control have already been described and include: ultra violet (UV) radiation [86]; direct electrolysis [87]; hydrodynamic cavitation [88]; ultrasound [89]; electron-beam radiation [90] and antimicrobial nanomaterials (e.g., chitosan, silver/aqueous fullerene nanoparticles, fullerol, photocatalytic titanium dioxide and carbon nanotubes) [91]. Some of these techniques are environmentally-friendly and less dangerous for health (e.g., electron-beam radiation, photocatalytic based technology, ultrasound). Sometimes, the use of two or more of the referred methods improves the disinfection efficiency. Also, combination of both chemical and non-chemical techniques can provide promising approaches for DW disinfection [3]. Indeed, some researchers have already demonstrated that the simultaneous application of two methods revealed more effectiveness in the eradication of DW biofilms than the two treatments applied separately [92, 93].

Based on the latest observations and because so many factors can influence biofilm growth, a more successful DW disinfection program should incorporate multi-faceted approaches to control established biofilms as well as to prevent future occurrences/regrowth events.

CONCLUSION

Water can be a means of diseases transmission posing great threats to global public health. This situation has increased the concern of the regulatory entities in providing to the populations access to clean and safe water. However, many microorganisms have demonstrated the ability to survive in DWDS during disinfection cycles and then proliferate. Besides, due to many favorable factors (e.g., nutrient levels, organic carbon) that exist within DWDS, some of them possess the ability to attach to the pipe wall and produce biofilms. These sessile communities confer to their cells a privileged environment where they are more protected from various stresses, such as disinfection steps.

The presence of biofilms in DWDS entails several problems, from aesthetics to pipes corrosion, which affect water quality. But the main drawback is their potential to serve as a protective niche for waterborne pathogens in which they can persist and even multiply. Further, biofilms, generally contribute to the continuous water contamination (without a contamination event) through the bacteria detachment of the pipe surfaces. Thus, as they act as a reservoir for coliforms and other pathogenic bacteria, the accuracy of the indicator functions of coliforms, such as *E. coli* is reduced. Moreover, the presence of some virulent *E. coli* strains is not contemplated by the routine microbiological parameters. This gap, associated with the complexity of the methodologies commonly used to detect and enumerate some naturally occurring *E. coli* serotypes, like *E. coli* O157:H7, exacerbates this problem. At the same time, this variants like other enteric bacteria can enter in a viable but non-cultivable state.

In this context, reliable and more sensitive/specific detection technologies non-culture based are essential for providing microbial safe DW. As, with conventional disinfection processes is almost impossible prevent the development of microbial biofilms and/or promote their eradication from DWDS, novel alternative disinfection strategies are urgently needed. Distinctive approaches have been proposed, but there are no method with reliable efficacy. Also, sometimes they are more expensive than chemical disinfection and no residual disinfectant is maintained in DWDS, requiring a second step. On the other hand, they are frequently more environmentally-friendly with less or no by-products production. In this way, the application of manifold approaches, including hybrid methods, can avoid the addition of high disinfectant concentrations and would greatly contribute to a safer DW with minimal impacts on environment and human health.

Acknowledgments

This work was financially supported by Project UID/EQU/00511/2013-LEPABE, by the FCT/MEC with national funds, and when applicable it is cofounded by FEDER in the scope of the P2020 Partnership Agreement; Project NORTE-07-0124-FEDER-000025-RL2_Environment & Health, by FEDER funds through Programa Operacional Factores de Competitividade (COMPETE), by the Programa Operacional do Norte (ON2), and by national funds through FCT (Fundação para a Ciência e a Tecnologia); by COMPETE, FCT/MEC (PIDDAC), and FEDER through Project Phytodisinfectants, PTDC/DTP-SAP/1078/2012 (COMPETE: FCOMP-01-0124-FEDER-028765), and the PD grant awarded to Anabela Borges (SFRH/BPD/98684/2013).

References

[1] Szewzyk U, Szewzyk R, Manz W, Schleifer KH (2000) Microbiogical safety of drinking water. *Annual Review of Microbiology* 54:81-127.

[2] WHO (2011) Guidelines for drinking-water quality, fourth edition. Geneva: World Health Organization.:1-564.

[3] Chaves Simoes L, Simoes M (2013) Biofilms in drinking water: problems and solutions. RSC Advances 3 (8):2520-2533.

[4] Dawson DJ, Sartory DP (2000) Microbiological safety of water. *British Medical Bulletin* 56 (1):74-83.

[5] WHO (2014) Water Safety in Distribution Systems.1-147.

[6] Yang K, LeJeune J, Alsdorf D, Lu B, Shum CK, Liang S (2012) Global distribution of outbreaks of water-associated infectious diseases. *PLoS Neglected Tropical Diseases* 6 (2).

[7] EPA (2002) United States Environmental Protection Agency. Health Risks from Microbial Growth and Biofilms in Drinking Water Distribution Systems.

[8] September SM, Els FA, Venter SN, Brözel VS (2007) Prevalence of bacterial pathogens in biofilms of drinking water distribution systems. *Journal of Water and Health* 5 (2):219-227.

[9] Berry D, Xi C, Raskin L (2006) Microbial ecology of drinking water distribution systems. *Current Opinion in Biotechnology* 17 (3):297-302.

[10] Blasi MF, Carere M, Pompa MG, Rizzuto E, Funari E (2008) Water-related diseases outbreaks reported in Italy. *Journal of Water and Health* 6 (3):423-432.

[11] Beaudeau P, De Valk H, Vaillant V, Mannschott C, Tillier C, Mouly D, Ledrans M (2008) Lessons learned from ten investigations of waterborne gastroenteritis outbreaks, France, 1998-2006. *Journal of Water and Health* 6 (4):491-503.

[12] Craun MF, Craun GF, Calderon RL, Beach MJ (2006) Waterborne outbreaks reported in the United States. *Journal of Water and Health* 4 (SUPPL. 2):19-30.

[13] Juhna T, Birzniece D, Larsson S, Zulenkovs D, Sharipo A, Azevedo NF, Ménard-Szczebara F, Castagnet S, Féliers C, Keevil CW (2007) Detection of *Escherichia coli* in biofilms from pipe samples and coupons in drinking water distribution networks. *Applied and Environmental Microbiology* 73 (22):7456-7464.

[14] Brettar I, Höfle MG (2008) Molecular assessment of bacterial pathogens — a contribution to drinking water safety. *Current Opinion in Biotechnology* 19 (3):274-280.

[15] Reynolds KA, Mena KD, Gerba CP (2008) Risk of waterborne illness via drinking water in the United States. *Rev Environ Contam Toxicol* 192:117-158.

[16] Hunter PR, Colford JM, LeChevallier MW, Binder S, Berger PS (2001) Waterborne diseases. Emerging Infectious *Diseases* 7 (3 Suppl):544-544.

[17] Cabral JPS (2010) Water Microbiology. Bacterial Pathogens and Water. *International Journal of Environmental Research and Public Health* 7 (10):3657-3703.

[18] CDC (2013) (Center for Disease Control and Prevention). Surveillance for waterborne disease outbreaks-United States. *MMWR* 62 (35):709-732.

[19] Ford TE (1999) Microbiological safety of drinking water: United States and global perspectives. *Environmental Health Perspectives* 107 (Suppl 1):191-206.

[20] WHO (2009) Europe, in Outbreaks of waterborne diseases, European Environment and Health Information System, (www.euro.who.int/ ENHIS).

[21] Ashbolt NJ (2004) Microbial contamination of drinking water and disease outcomes in developing regions. *Toxicology* 198 (1–3):229-238.

[22] Schoenen D (2002) Role of disinfection in suppressing the spread of pathogens with drinking water: possibilities and limitations. *Water Research* 36 (15):3874-3888.

[23] Gomes IB, Simões M, Simões LC (2014) An overview on the reactors to study drinking water biofilms. *Water Research* 62 (0):63-87.

[24] Schwering M, Song J, Louie M, Turner RJ, Ceri H (2013) Multi-species biofilms defined from drinking water microorganisms provide increased protection against chlorine disinfection. *Biofouling* 29 (8):917-928.

[25] Percival SL, Walker JT (1999) Biofilms and public health significance. *Biofouling* 14:99-115.

[26] Stoodley P, Sauer K, Davies DG, Costerton JW (2002) Biofilms as complex differentiated communities. *Annual Review of Microbiology* 56:187-209.

[27] Simões M (2011) Antimicrobial strategies effective against infectious bacterial biofilms. *Current Medicinal Chemistry* 18 (14):2129-2145.

[28] Costerton JW, Lewandowski Z, Caldwell DE, Korber DR, Lappin-Scott HM (1995) Microbial biofilms. *Annual Review of Microbiology* 49:711-745.

[29] Cos P, Toté K, Horemans T, Maes L (2010) Biofilms: An extra hurdle for effective antimicrobial therapy. *Curr Pharm Design* 16 (20):2279-2295.

[30] Camper AK (2004) Involvement of humic substances in regrowth. *International Journal of Food Microbiology* 92 (3):355-364.

[31] Beech IB (2004) Corrosion of technical materials in the presence of biofilms—current understanding and state-of-the art methods of study. *International Biodeterioration & Biodegradation* 53 (3):177-183.

[32] Teng F, Guan YT, Zhu WP (2008) Effect of biofilm on cast iron pipe corrosion in drinking water distribution system: corrosion scales characterization and microbial community structure investigation. *Corrosion Science* 50 (10):2816-2823.

[33] Emtiazi F, Schwartz T, Marten SM, Krolla-Sidenstein P, Obst U (2004) Investigation of natural biofilms formed during the production of drinking water from surface water embankment filtration. *Water Research* 38 (5):1197-1206.

[34] Lührig K, Canbäck B, Paul CJ, Johansson T, Persson KM, Rådström P (2015) Bacterial community analysis of drinking water biofilms in southern Sweden. *Microbes and Environments* 30 (1):99-107.

[35] Camper A, Burr M, Ellis B, Butterfield P, Abernathy C (1998) Development and structure of drinking water biofilms and techniques for their study. *Journal of Applied Microbiology* 85 (S1):1S-12S.

[36] Flemming HC, Percival SL, Walker JT (2002) Contamination potential of biofilms in water distribution systems. *Water Science and Technology: Water Supply,* vol 2.

[37] Huq A, Whitehouse CA, Grim CJ, Alam M, Colwell RR (2008) Biofilms in water, its role and impact in human disease transmission. *Current Opinion in Biotechnology* 19 (3):244-247.

[38] Faruque SM, Biswas K, Udden SMN, Ahmad QS, Sack DA, Nair GB, Mekalanos JJ (2006) Transmissibility of cholera: in vivo-formed biofilms and their relationship to infectivity and persistence in the environment. *Proceedings of the National Academy of Sciences of the United States of America* 103 (16):6350-6355.

[39] Kilb B, Lange B, Schaule G, Flemming H-C, Wingender J (2003) Contamination of drinking water by coliforms from biofilms grown on rubber-coated valves. *International Journal of Hygiene and Environmental Health* 206 (6):563-573.

[40] LeChevallier MW (1990) Coliform regrowth in drinking water: a review. *Journal / American Water Works Association* 82 (11):74-86.

[41] LeChevallier MW, Cawthon CD, Lee RG (1988) Inactivation of biofilm bacteria. *Applied and Environmental Microbiology* 54 (10):2492-2499.

[42] Keevil CW, Mackerness CW, Colbourne JS (1990) Biocide treatment of biofilms. *International Biodeterioration* 26 (2–4):169-179.

[43] Jefferson KK (2004) What drives bacteria to produce a biofilm? *FEMS Microbiological Letters* 236 (2):163-173.

[44] Johnson LR (2008) Microcolony and biofilm formation as a survival strategy for bacteria. *Journal of Theoretical Biology* 251 (1):24-34.

[45] Parsek MR, Singh PK (2003) Bacterial biofilms: an emerging link to disease pathogenesis. *Annual Review of Microbiology* 57:677-701.

[46] Hall-Stoodley L, Costerton JW, Stoodley P (2004) Bacterial biofilms: from the natural environment to infectious diseases. *Nature Reviews Microbiology* 2 (2):95-108.

[47] Elvers KT, Leeming K, Lappin-Scott HM (2002) Binary and mixed population biofilms: time-lapse image analysis and disinfection with biocides. *Journal of Industrial Microbiology and Biotechnology* 29 (6):331-338.

[48] Kart D, Tavernier S, Van Acker H, Nelis HJ, Coenye T (2014) Activity of disinfectants against multispecies biofilms formed by *Staphylococcus aureus*, *Candida albicans* and *Pseudomonas aeruginosa*. *Biofouling* 30 (3):377-383.

[49] Simões LC, Simões M, Vieira MJ (2010) Influence of the diversity of bacterial isolates from drinking water on resistance of biofilms to disinfection. *Applied and Environmental Microbiology* 76 (19):6673-6679.

[50] De Beer D, Srinivasan R, Stewart PS (1994) Direct measurement of chlorine penetration into biofilms during disinfection. *J Appl Environ Microbiol* 60 (12):4339-4344.

[51] Gilbert P, McBain AJ, Rickard AH (2003) Formation of microbial biofilm in hygienic situations: A problem of control. *International Biodeterioration and Biodegradation* 51 (4):245-248.

[52] Anderson GG, O'Toole GA (2008) Innate and induced resistance mechanisms of bacterial biofilms. *Curr Top Microbiol Immunol* 322:85-105.

[53] Stewart PS (2002) Mechanisms of antibiotic resistance in bacterial biofilms. *International Journal of Medical Microbiology* 292 (2):107-113.

[54] Falkinham Iii JO, Norton CD, Lechevallier MW (2001) Factors influencing numbers of *Mycobacterium avium*, *Mycobacterium intracellulare*, and other Mycobacteria in drinking water distribution systems. *J Appl Environ Microbiol* 67 (3):1225-1231.

[55] Watson CL, Owen RJ, Said B, Lai S, Lee JV, Surman-Lee S, Nichols G (2004) Detection of *Helicobacter pylori* by PCR but not culture in water and biofilm samples from drinking water distribution systems in England. *Journal of Applied Microbiology* 97 (4):690-698.

[56] Costa J, Tiago I, da Costa MS, Veríssimo A (2005) Presence and persistence of *Legionella* spp. in groundwater. *J Appl Environ Microbiol* 71 (2):663-671.

[57] Flemming HC (2002) Biofouling in water systems – cases, causes and countermeasures. *Applied Microbiology and Biotechnology* 59 (6):629-640.

[58] Percival S, Chalmers R, Embrey M, Hunter P, Sellwood J, Wyn-Jones P (2004) Chapter Seven - *Escherichia coli*. In: *Wyn-Jones SPCEHS (ed) Microbiology of Waterborne Diseases*. First Edition edn. Academic Press, London, pp 71-90.

[59] Leclerc H, Mossel DAA, Edberg SC, Struijk CB (2001) Advances in the bacteriology of the coliform group: their suitability as markers of microbial aater safety. *Annual Review of Microbiology* 55 (1):201-234.

[60] Percival SL, Williams DW (2014) Chapter Six - *Escherichia coli*. In: *Gray SLPVYWWMCF (ed) Microbiology of Waterborne Diseases*. Second Edition edn. Academic Press, London, pp 89-117.

[61] Muniesa M, Jofre J, García-Aljaro C, Blanch AR (2006) Occurrence of *Escherichia coli* O157:H7 and other enterohemorrhagic *Escherichia coli* in the environment. *Environmental Science and Technology* 40 (23):7141-7149.

[62] Schroeder S, Caldwell J, Vernon T, White P, Granger S, Bennett J (1968) A waterborne outbreak of gastroenteritis in adults associated with *Escherichia coli*. *The Lancet* 291 (7545):737-740.

[63] Rosenberg ML, Koplan JP, Wachsmuth IK, Wells JG, Gangarosa EJ, Guerrant RL, Sack DA (1977) Epidemie diarrhea at Crater Lake from enterotoxigenic *Escherichia coli* a large waterborne outbreak. *Annals of Internal Medicine* 86 (6):714-718.

[64] Kudoh Y, Zen-Yoji H, Matsushita S, Sakai S, Maruyama T (1977) Outbreaks of acute enteritis due to heat-stable enterotoxin-producing strains of *Escherichia coli*. *Microbiology and Immunology* 21 (3):175-178.

[65] Bavaro M (2012) *E. coli* O157:H7 and other toxigenic strains: the curse of global food distribution. *Curr Gastroenterol Rep* 14 (4):317-323.

[66] Leclerc H, Schwartzbrod L, Dei-Cas E (2002) Microbial agents associated with waterborne diseases. *Critical Reviews in Microbiology* 28 (4):371-409.

[67] Keene WE, McAnulty JM, Hoesly FC, Williams LP, Hedberg K, Oxman GL, Barrett TJ, Pfaller MA, Fleming DW (1994) A swimming-associated outbreak of hemorrhagic colitis caused by *Escherichia coli* O157:H7 and *Shigella Sonnei*. *New England Journal of Medicine* 331 (9):579-584.

[68] Isaäcson M, Canter PH, Effler P, Arntzen L, Bomans P, Heenan R (1993) Haemorrhagic colitis epidemic in Africa. *The Lancet* 341 (8850):961.

[69] Rangel J, M., Sparling P, H., Crowe C, Griffin P, M., Swerdlow D, L. (2005) Epidemiology of *Escherichia coli* O157:H7 Outbreaks, United States, 1982–2002. *Emerg Infect Dis* 11 (4):603.

[70] Oliver JD (2010) Recent findings on the viable but nonculturable state in pathogenic bacteria. *FEMS Microbiol Rev* 34 (4):415-425.

[71] Trevors JT (2011) Viable but non-culturable (VBNC) bacteria: gene expression in planktonic and biofilm cells. *Journal of Microbiological Methods* 86 (2):266-273.

[72] Gould LH (2012) Update: Recommendations for Diagnosis of shiga toxin-producing *Escherichia coli* infections by clinical laboratories. *Clinical Microbiology Newsletter* 34 (10):75-83.

[73] Camper AK, McFeters GA, Characklis WG, Jones WL (1991) Growth kinetics of coliform bacteria under conditions relevant to drinking water distribution systems. *J Appl Environ Microbiol* 57 (8):2233-2239.

[74] LeChevallier MW, Welch NJ, Smith DB (1996) Full-scale studies of factors related to coliform regrowth in drinking water. *Applied and Environmental Microbiology* 62 (7):2201-2211.

[75] Fass S, Dincher ML, Reasoner DJ, Gatel D, Block JC (1996) Fate of *Escherichia coli* experimentally injected in a drinking water distribution pilot system. *Water Research* 30 (9):2215-2221.

[76] Williams MM, Braun-Howland EB (2003) Growth of *Escherichia coli* in model distribution system biofilms exposed to hypochlorous acid or monochloramine. *Applied and Environmental Microbiology* 69 (9):5463-5471.

[77] Weber H, Polen T, Heuveling J, Wendisch VF, Hengge R (2005) Genome-wide analysis of the general stress response network in *Escherichia coli*: σ(S)-dependent genes, promoters, and sigma factor selectivity. *Journal of Bacteriology* 187 (5):1591-1603.

[78] Bauman WJ, Nocker A, Jones WL, Camper AK (2009) Retention of a model pathogen in a porous media biofilm. *Biofouling* 25 (3):229-240.
[79] Moreira S, Brown A, Ha R, Iserhoff K, Yim M, Yang J, Liao B, Pszczolko E, Qin W, Leung KT (2012) Persistence of *Escherichia coli* in freshwater periphyton: biofilm-forming capacity as a selective advantage. *FEMS Microbiol Ecology* 79 (3):608-618.
[80] Ahmed D, Islam MS, Begum YA, Janzon A, Qadri F, Sjöling Å (2013) Presence of enterotoxigenic *Escherichia coli* in biofilms formed in water containers in poor households coincides with epidemic seasons in Dhaka. *Journal of Applied Microbiology* 114 (4):1223-1229.
[81] Farkas A, Drăgan-Bularda M, Ciatarâş D, Bocoş B, Ţigan Ş (2012) Opportunistic pathogens and faecal indicators in drinking water associated biofilms in Cluj, Romania. *Journal of Water and Health* 10 (3):471-483.
[82] Deborde M, von Gunten U (2008) Reactions of chlorine with inorganic and organic compounds during water treatment—kinetics and mechanisms: a critical review. *Water Research* 42 (1–2):13-51.
[83] Zhou L-l, Zhang Y-j, Li G-b (2009) Effect of pipe material and low level disinfectants on biofilm development in a simulated drinking water distribution system. *J Zhejiang Univ Sci A* 10 (5):725-731.
[84] Friedman M, Kirmeyer GJ, Antoun E (2002) Developing and implementing a distribution system flushing program. *Journal / American Water Works Association* 94 (7):48-56.
[85] Antoun EN, Dyksen J, Hiltebrand DJ (1999) Unidirrectional flushing: a powerful tool. *Journal / American Water Works Association* 91 (7):62-71.
[86] Hijnen WAM, Beerendonk EF, Medema GJ (2006) Inactivation credit of UV radiation for viruses, bacteria and protozoan (oo)cysts in water: a review. *Water Research* 40 (1):3-22.
[87] Martínez-Huitle CA, Brillas E (2008) Electrochemical alternatives for drinking water disinfection. *Angewandte Chemie International Edition* 47 (11):1998-2005.
[88] Mezule L, Tsyfansky S, Yakushevich V, Juhna T (2009) A simple technique for water disinfection with hydrodynamic cavitation: effect on survival of *Escherichia coli*. *Desalination* 248 (1–3):152-159.
[89] Piyasena P, Mohareb E, McKellar RC (2003) Inactivation of microbes using ultrasound: a review. *International Journal of Food Microbiology* 87 (3):207-216.

[90] Strokin NA (2007) Electron-beam treatment of drinking water in an industrial system. *High Energy Chem* 41 (1):1-4.

[91] Li Q, Mahendra S, Lyon DY, Brunet L, Liga MV, Li D, Alvarez PJJ (2008) Antimicrobial nanomaterials for water disinfection and microbial control: potential applications and implications. *Water Research* 42 (18):4591-4602.

[92] Rand JL, Hofmann R, Alam MZB, Chauret C, Cantwell R, Andrews RC, Gagnon GA (2007) A field study evaluation for mitigating biofouling with chlorine dioxide or chlorine integrated with UV disinfection. *Water Research* 41 (9):1939-1948.

[93] Murphy HM, Payne SJ, Gagnon GA (2008) Sequential UV- and chlorine-based disinfection to mitigate *Escherichia coli* in drinking water biofilms. *Water Research* 42 (8–9):2083-2092.

In: Coliforms
Editor: Geoffrey McCoy

ISBN: 978-1-63483-542-8

Chapter 2

METHODS OF ANALYSIS FOR BACTERIAL CONTAMINATION IN ENVIRONMENTAL WATERS

Robert G. Price[1] and Dirk Wildeboer[2]
[1] King's College London, Faculty of Life Sciences and Medicine, Diabetes and Nutritional Sciences Division, Franklin-Wilkins Building, London, UK
[2] Middlesex University, School of Science and Technology, Natural Sciences Department, The Burroughs, London, UK

ABSTRACT

Water pollution by faecal contamination is a serious problem due to the potential for contracting disease if the water is consumed untreated. Coliform bacteria *Escherichia coli* and *Enterococci* are used as primary indicators of contamination in fresh and marine water. The presence and levels of these organisms are indicative of the presence of a number of pathogenic organisms of animal origin and thus for a potential threat to human health. The acceptable levels of indicator organisms are defined by legislation and vary for drinking, river, well and marine water.

This review will consider currently employed methods of analysis for coliform bacteria, compare them with recently developed procedures and evaluate advantages and limitations of different methods. Emphasis will be placed on methods which can be used on-site with laboratory procedures as confirmatory procedures. Established culture methods use chromogenic substrates for β-D-glucuronidase to identify *E. coli* and β-D-galactosidase to detect *Enterococci*. Rapid non-culture based fluorimetric procedures have been developed that can be used on-site. In addition to culture and biochemical procedures, advances have been made in developing genomic, proteomic and metabolomic methods. Methods must be quantitative in order to comply with legal norms and in this regard the potential involvement of biosensor technology will be of great value in successfully transferring laboratory procedures to the field.

The Chapter will conclude with a recommendation of the most valuable laboratory and on-site methods for water analysis given the constraints of the current legal framework governing acceptable water quality.

Introduction

Contamination of environmental water is a is a serious problem affecting rivers, sea beaches, lakes, ground water, surface water, recreational water and the recreational activity associated with these sites (Scholes, 2008). Microbiological contamination of rivers and beaches can result from leakage of sewage, water draining from land and from animals and birds (EEA, 2013). Contamination of rivers (Hall, 2013) and beaches (Boehm, 2014) can result in outbreak of gastroenteritis, even low contact is significant but the amount of water ingested is most important when determining risk (Dorevitch, 2012). The health risk can be assessed (monitored) using a series of tests (indicators) which are defined in regional legislation (Table 1). The United States Environmental Protection agency (US EPA, 2012) criteria are largely based on a prospective survey design and included molecular methods for bacterial enumeration, while World Health Organisation (WHO, 2003) guidelines and European Union (EU, 2006) standards are based on a randomised control approach with culture methods to define microbial exposure (Fewtrell, 2015).

Faecal microorganisms enter environmental water via waste water, sewage release, surface and subsurface run-off. The acceptable upper limit for the presence of *E. coli* set by the EU is 900 cfu/100 ml in freshwater and 500 cfu/100 ml in marine water (EU, 2006). In the same directive the upper limit of *enterococci* is 330 cfu/100ml for inland water and 185 cfu/100ml for coastal waters (Table 1).

Table 1. Legal levels defined for bacterial contamination of environmental waters

Legislative Body	Organism	Test principle	Type of water body	Tolerable maximum (cfu/100 ml)	Reference
WHO	*Enterococci*	"standardised method"	Recreational water (coastal and fresh)	500	WHO (2003)
US EPA	*Enterococci*	EPA Method 1600	Marine and fresh	130	US EPA (2012), US EPA (2002a)
US EPA	*E. coli*	EPA Method 1603	Fresh	410	US EPA (2012), US EPA (2002b)
European Union	*Enterococci*	Culture, MPN	Costal/Estuarine Freshwater	185 330	EU (2006)
European Union	*E. coli*	Culture, MPN	Costal/Estuarine Inland	500 900	EU (2006)
UK (England & Wales)	*E. coli*	Culture, MPN	Bathing (coastal and inland)	2000	UK EA (2015)
Australia	*Enterococci*	reference methods	Recreational water (coastal and fresh)	500	AUS Gov (2008)
Brazil	Coliforms *Enterococci* *E. coli*	Culture, MPN	Recreational Marine only Recreational	1000 100 800	CONAMA (2000)
South Africa	*Enterococci* *E. coli*	"Standard Methods for the Examination of Water and Wastewater"	Coastal/Marine	185 500	SA EA (2012)
South Africa	Coliforms *Enterococci* *E. coli*	Culture	Fresh	2000 100 400	SA EA (1996)

A number of reports have shown that most *E. coli* (over 95%) express β-D-glucuronidase (GUD) activity (Manafi, 1991) making this enzyme a sensitive and specific marker for *E. coli* and thus faecal pollution. Environmental monitoring of bacterial content is essential to ensure compliance within the legal ranges and is dependent on established and recently developed tests.

Ideally tests should give a result on-site with confirmation later in the laboratory. However the currently agreed tests depend on culture procedures and results can take 24 to 48 h to complete. Molecular methods are used to determine the total bacterial content of contaminated water. The Coliform group are proven indicators of faecal contamination (OECD, 2004). Although they don't usually cause serious illness their presence is used to indicate the presence of other pathogenic bacteria and viruses including *Norovirus* (Godoy, 2006; Lodder, 2005). Coliforms are easy to cultivate and β-D-galactosidase has been used as an indicator of the presence of these organisms in water (Tallon, 2005). More recently β-D-glucuronidase is widely used as the main molecular marker as it is a better indicator of the presence of *E. coli* (Ryzinska-Paier, 2012). *E. coli* is widely accepted as the better faecal indicator organism than total Coliforms. Other organisms of interest include *Streptococci* (temperate and marine waters), *Enterococci* (marine waters), *Salmonella* and *Clostridium perfringens* are listed in Table 2.

Table 2. Target microorganisms and diagnostic enzymes

Microorganism	Diagnostic Enzyme	Reference
E. coli	β-D-glucuronidase	Cassar and Cuschieri (2003)
Coliforms	β-D-galactosidase	Browne et al., (2010)
Enterococci	β-D-glucosidase	US EPA (2002)
Candida	Hexosaminidase / Alkaline phosphatase	Cooke (2002), Ibrahim (2001)
Staphylococci	Coagulase / Mononuclease	Victor (1969)
Enterobacter	α-glucosidase	Song (2008)
Proteus	Tryptophan deaminase	Samra (1998)
E coli 0157:H7	β-D-glucuronidase negative	Perry (2003)
Salmonella	α-galactosidase; C8 Ester	Perez (2003), Cooke (1999)
Clostridium perfringens	Acid phosphatase/β-D-galactosidase	Adcock and Saint (2001)

Note: Media may incorporate fluorogenic or chromogenic substrates for the diagnostic enzymes.

Culture methods for the enumeration of indicator bacteria are time consuming and rely on biochemical or immunological characterisation. However, for many well-known and emerging pathogens culture methods and biochemical follow-up procedures are not available (Dong, 2008). Moreover culture methods may underestimate bacterial load. Alternative new technologies principally nucleic acid based methods and their application to environmental samples have been reviewed (Fewtrell, 2015; Girones, 2010; Call, 2003; OECD, 2004). The methods used must be quantitative in order to comply with legal norms and in this regard the potential involvement of biosensor technology will be of great value in successfully transferring laboratory procedures to the field.

A key problem to be addressed is whether a test measures all the bacteria present including the non-viable organisms (Nebe-von-Caron, 2000; OECD, 2004). After prolonged exposure to water, bacterial pathogens can enter a viable but non-culturable stage, in which they are viable but cannot be detected by culture techniques (Cenciarini-Borde, 2009). Further bacteria may be embedded in biofilms which are not accessible by the established laboratory procedures. Culture procedures measure only the viable organisms present in the samples. However a significant number of cells may not be detected despite being viable and they are classified as viable but non-culturable cells (VBNC). This may be due to the effect of for example stress (Edwards, 1996), the type of medium or physical conditions in the sample, e.g., oxygen, nutrient content of the sample (McPeters, 1982). Chromogenic agar can detect non–growing cells by measuring the presence of an enzyme (Table 2). If there is damage to DNA the cell may not be metabolically active however if its membrane is unaffected it is still considered viable. Dead or permeated cells readily take up DNA stains and are classified as non-viable. For these reasons culture techniques may underestimate the viable bacterial load in a sample, while techniques that also measure metabolically active but not growing or cells with full membrane integrity would give higher numbers of cells than the culture procedures.

A tiered system of analysis methods is the most successful approach to environmental water analysis (see Figure 1). Screening methods can be qualitative or semi-qualitative but should be capable of delivering accurate results within twenty-four hours. Some automated systems based on the detection of β-D-glucuronidase are available (Tryland, 2002) but for bathing waters a more rapid procedure would be valuable (Amirat, 2012). The second tier involves enumeration which involves the quantification of bacteria present.

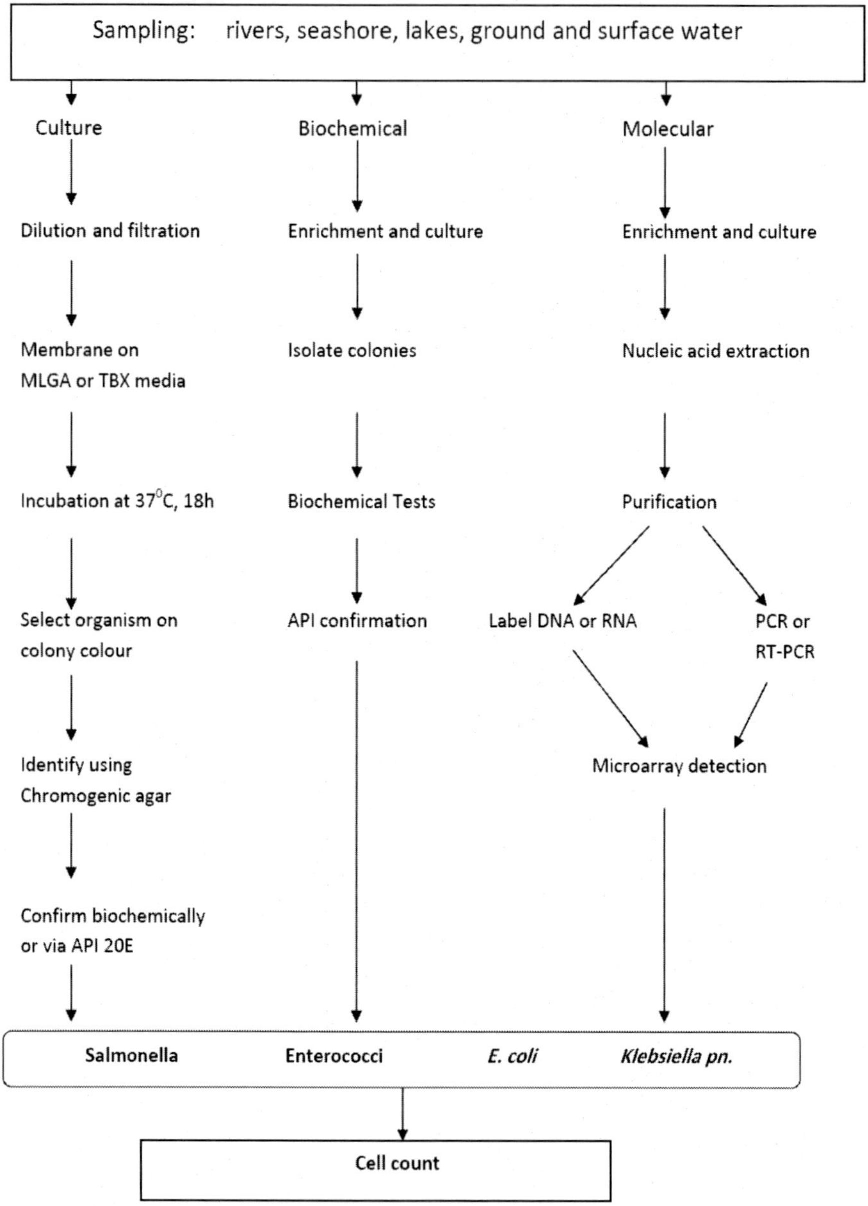

Figure 1. Flow chart showing the different steps followed for isolation, detection, identification and confirmation of key bacteria from environmental water. MPN procedure is not shown.

The third tier is the identification and characterisation of the type of microorganisms present. There are a wide range of methods available to characterise the type of bacteria present including chromogenic media. Molecular methods can be employed to determine the source of contamination e.g., whether it is from human or animal sources. In this review the tests used to define the legally accepted ranges of bacteria with be discussed followed by a critical assessment of the more recent tests emphasising those which have potential to be used on-site.

CULTURE BASED METHODS

Water quality is based on the number of indicator organisms in a given volume of water and is carried out in selective media (Browne, 2010). The presence of *Escherichia coli* is now accepted as the best indicator of faecal contamination. Some strains of *E. coli* which are β-D-glucuronidase negative, such as *Escherichia coli* O157:H3, will not be detected as *E. coli* but, as they are β-D-galactosidase positive, they will appear as coliform bacteria on selected chromogenic agars (Heijnen, 2006). Enzyme activity can be detected either using chromogenic, fluorogenic or bioluminescent substrates. A number of automated systems are currently available namely Colifast Analyser (Colifast, Norway) based on 4-methylumbelliferone-β-D-galactosidase is used for on-line monitoring and is now available in a micro hand held version. The Colilert® 3000 system (Seres, France) utilises fluorescent or chromogenic substrates and can deliver results within twenty-four hours. This method correlated well with standard laboratory methods although the results were two to three orders of magnitude higher than membrane filter and MPN methods probably due to the presence of *Aeromonas spp.* and *Vibrio spp.* (natural inhabitants of the raw surface water) known to interfere with the Colilert test (Zuckerman, 2008). Results were obtained in 6-10 h rather than the 24 hours required by membrane filtration tests. These systems can be modified to measure other organisms by changing the enzyme substrate, e.g., the fluorescent glucoside to detect *Enterococci,* or conditions.

The reference methods for detection and isolation of *E. coli* and Coliforms from water are the membrane filtration method (ISO 9308-1:2014) and the fluorescent miniaturised most probable number (MPN, ISO 9308-2:2012). ISO 9308-1:2014 is based on membrane filtration and subsequent culture on a chromogenic coliform agar medium (ISO, 2015). Due to the low selectivity of the differential agar medium, background growth can interfere with the reliable enumeration of *E. coli* and coliform bacteria, for example, in surface waters or shallow well waters. This method is not suitable for these types of water. ISO 9308-1:2014 is especially suitable for waters with low bacterial numbers that will cause less than 100 total colonies on chromogenic coliform agar (CCA). These may be drinking water, disinfected pool water, or finished water from drinking water treatment plants. As the MPN method (ISO 9308-2:2012) is based on the growth of the target organisms in liquid medium it is suitable for most water but should not be used for enumeration of bacteria in marine samples as dilution of the sample is required. Amirat et al., (2012) have shown that the membrane filtration followed by culture on selective chromogenic media can be used to monitor bacterial contamination of the river Thames. In this study the microbial quality of the river Thames was investigated at high and low tide and in different seasons with the aim of demonstrating whether the levels fell within the accepted guidelines for bathing water. Water was separated from any sediment prior to dilution with Ringers solution followed by filtration through a 0.45 μm membrane and incubation on MLGA or TBX agar media for up to 18 hours. Microorganism types were separated on the basis of colour and cfu/100ml calculated. Further identification was carried out by application to defied media or chemical tests. This procedure successfully identified *Salmonella*, *Enterococci*, *Klebsiella pneumoniae* and *E. coli* (Figure 1). Sixty percent of the samples were in excess of the EU standard for bathing water and the study demonstrated frequent sewage pollution of the Thames which was most noticeable after heavy rainfall. These results confirmed those of an earlier study by Tryland et al., (2002) who used the Colifast Early Warning system based on the fluorimetric detection of β-D-glucuronidase. Kolarevic et al., (2011) studied the river Tisa in India using the membrane filtration method followed by culture for faecal coliforms and intestinal *Enterococci*. The results were confirmed using the API 20e identification kit (Biomerieux) procedures to determine sanitary quality of the river. To accomplish the sanitary analyses, the presence of *Proteus sp.*, *Pseudomonas aeruginosa* and *Bacillus sp.* were cultivated on specialised media. In a study of the river Göta Älv in Sweden Tornevi et al., (2014) used the MPN method to demonstrate an increase in indicator organisms including *E. coli* two days after rainfall. Skariyachan et al., (2015) also used MPN and biochemical characterisation to measure faecal pollution and antibiotic resistance in the river Cauvery in India. The faecal contamination of the river Danube was measured using the indicator organisms Coliforms and *Enterococci* (Pall, 2013).

A wide range of media are available for the characterisation of environmental microorganisms (Atlas, 2010) and won't be discussed further here. Detailed descriptions of standard laboratory procedures which are used in environmental studies including microscopic as well as biochemical characterisation are given in Alexander and Strete (2001). The API® ID Strip Range (BioMerieux, France) allows the fast identification of known bacteria and consists of a series of miniaturised techniques based on established laboratory procedures.

MOLECULAR ANALYSIS

To detect a particular organism in an environmental a sample it must have at least one unique trait or nucleotide sequence which is stable under the physiological conditions encountered in environmental samples. This trait allows it to be distinguished from the other bacteria present. Secondly there should be a robust detection method available that allows the (semi-) quantitative assessment of the trait (OECD, 2004). Viable organisms can be detected by culture based methods, fluorescent tags, immuno-based methods as well as those based on RNA or DNA analysis. Viable but non-culturable (VBNC) organisms can be detected by all the procedures except culture, ghost or lysed cells can only be determined by immuno-based techniques while cell debris can only be assessed using DNA based techniques.

Detection of marker gene products which are specific for an organism may be detected by measurement of an enzyme activity or on the basis of its physico-chemical properties. DNA and RNA hybridisation methods rely on the assessment of DNA sequence of a marker gene that is specific for the marker organism or of RNA transcribed for the sequence. Detection methods are based on hybridisation of probes, tagged for detection to a specific target sequence. Hybridisation may be performed *in situ* detection is by either a fluorescence or chromogenic enzyme reaction, or radiolabelling. RNA is less stable than DNA and relies on active transcription of the sequence. Amplification of DNA or RNA sequences by the polymerase chain reaction (PCR) increases their sensitivity and PCR has the further advantage that it can be carried out quantitatively.

The coupling of microarray technology with PCR enhances product detection and identification (Call, 2003). The sensitivity of this procedure is limited by the amount of nucleic acid available and may lack the sensitivity required for environmental screening. When the amount of nucleic acid is not limiting, microarrays may prove valuable as a fingerprinting tool. Nucleic acid amplification procedures provide sensitive, rapid and quantitative analytical tools for detecting specific pathogens, including emerging strains and pathogens (Girones, 2010). DNA hybridisation techniques are costly and labour intensive and have been replaced by lower cost PCR based procedures, RT-PCR is rapid but there is a need to reduce the enrichment time required prior to the assay (Beneduce, 2003).

Molecular techniques for the specific detection and quantification of bacteria are highly sensitive, rapid and specific, they can be readily automated and standardised so have some advantages over the standard culture based techniques. However where low numbers of bacteria are present an enrichment step is required which does lengthen the time needed to complete the procedure (Noble, 2005). Quantitative (qPCR) analysis of the DNA in water samples can be carried out by the Taq-Man or SYBR green procedures to determine the number of genomes per volume of water for a specific bacterial species; details of concentrations recovered are given in Girones et al., (2010). This can be compared with the number of cells which may be determined by fluorescent *in situ* hybridisation (FISH), immunofluorescent microscopy or by simply determining the CFU by plate counting. Since bacterial cells can aggregate and may have more than one genome conversion factors need to be determined for each organism (Brettar, 2008). The FISH technique, based on hybridization rRNA oligonucleotide probes can be used to distinguish mixtures of microorganisms in environmental samples and study biofilms (Bottari, 2006; Juhna, 2007) allowing phylogenetic identification of mixtures of bacteria by epi-fluorescence and confocal microscopy. FISH procedures can also be carried out with DNA probes and oligonucleotides targeted to mRNA (Pernthaler, 2007). However the FISH procedure lacks sensitivity and usually requires a prior enrichment step.

Although the indicator organisms are faecal coliforms, *Streptococci* and *Enterococci* are more numerous in the Environment and *Salmonella* and *Campylobacter* are the most frequent causes of bacterial gastroenteritis (Westrell, 2009). *E. coli* O157:H7 is one of a group of emerging pathogens which are transmitted via the faecal route (Bavaro, 2009) and is responsible for local outbreaks of gastroenteritis. It has been identified in raw sludge, however the stability of this organism in sewage and waste water is currently unknown (Sidhu, 2009). The detection of these pathogenic organisms largely requires molecular methods. Several commercial kits are now available for the assay of *E. coli* O157:H7 in environmental samples but their primary use is in the analysis of suspected contaminated foods.

qPCR and qRT-PCR are rapid, sensitive techniques which provide information on the content of emerging and classical pathogens in environmental water samples, especially where culturing is unavailable. These procedures can be used to identify sources of contamination. However, these techniques do not provide information on the viability of bacteria and the associated risk. These methods still need standardisation and they are affected by the presence of inhibitors in the sample which affect extraction of the nucleic acid. Molecular methods provide an accurate guide to the bacterial load and can therefore be used in risk assessment. In a recent survey of recreational water and infection Fewtrell and Kay (2015) reviewed quantitative microbial risk assessment of infection (QMRAs) for recreational bathers. Since culture procedures can take over 24 h to complete molecular methods particularly qPCR are often used in tandem for indicator analysis in epidemiological studies (Arnold, 2013; Fleisher, 2010; Abdel Zaher, 2011; Colford, 2007; Cordero, 2012). With the technical advances and increased computing power that arrived with next-generation sequencing (NGS), whole genome sequencing (WGS) has grown in importance for the identification of bacteria in a complex sample. Recent studies have shown that this technology allows to identify the origin of faecal pollution in environmental waters (Gomi, 2014) and has the potential to detect alternative faecal pollution indicators (McLellan, 2014). As expected, when results were compared between culture and molecular methods, for example when *Enterococci* were assessed, the molecular methods gave higher results.

Table 3. Comparison of sensitivity and reliability of methods used to detect and characterise microorganism in environmental samples

Method	Technical Requirement	Detect	Time	sensitivity	Reliability
Culture – selective	enrichment (membrane filtration), selective base media	Viable cells only	24-48h	10^2 bacteria	Limited to organisms viable in the laboratory
Culture – chromogenic	enrichment (membrane filtration), chromogenic agar	Viable cells only	24-48h	10^2 bacteria 93%	Limited to organisms viable in the laboratory

Table 3. (Continued)

Method	Technical Requirement	Detect	Time	sensitivity	Reliability
MPN methods	serial dilution, specific media and reader	Viable cells only	6-18 h	2-300 cells	High
Biochemical – metabolites	Can be carried out on culture plates or in tubes (e.g., API system™)	Viable cells only	30 min to 24 h	Mostly qualitative	high, confirmatory tests
Biochemical – enzymes	Extraction of enzyme	Viable cells and VBNC	20 min to 12 h	High	Relies on constitutive expression and stability of gene product
DNA/RNA direct	Preparation of appropriate probes	Viable cells and VBNC	Slow, sample preparation and hybridisation time	Depends on probes and control of the hybridisation conditions	Depends on specificity of probes used, quantitative reliability is low
DNA/RNA amplification	Nucleotide extraction/ enrichment	Viable cells and VBNC, cell debris	Slow, sample preparation and amplification time	Up to a few DNA or RNA molecules	Care required to avoid false positives, sample might contain PCR inhibitors
Biosensor – biochemical	Enrichment might be required	Viable cells and VBNC	10 min – 2 h	<10 cfu/ml	Depends on bio- and signalling element
Biosensor – molecular	Enrichment might be required	Viable cells and VBNC, cell debris	<30 min	<10 cfu/ml	Depends on bio- and signalling element

Fewtrell and Kay (2015) provide an excellent summary of recent studies on monitoring infection and concluded that there is still a need to re-examine the precision and reproducibility of both the culture and molecular procedures and how they are affected through common sewage treatment. This is required to allow regulatory authorities to make an objective choice between the various analytical systems available. The sensitivity and reliability of the various methods used to estimate bacterial load in environmental samples are shown in Table 3.

BIOSENSORS

A biosensor is defined as a compact analytical device incorporating a biological or biologically-derived sensing element either integrated in or closely associated with a physiochemical transducer that produces either a discrete or continuous digital electronic signal proportional to the single analyte or a related group of analytes. The advantage of a biosensor technology is its capacity to directly detect the target and provide a result in real-time. In the context of this chapter, this means an output that directly informs on the level of bacterial contaminants present in environmental water. This can achieve a considerable time saving compared to culture based methods or protocols requiring extraction and purification of bacteria or molecules. Advances in electronic and optics engineering have allowed for smaller, low powered devices to be designed. Modern biosensors are portable and easy to use, making them ideal for on-site testing. Biosensors have been developed for a range of bacterial pathogens and their toxins, mostly for food-safety testing, including drinking water, and point-of-care testing in health care. The different biosensor technologies developed for selected waterborne pathogens are summarised by Xu and Mutharasan (2011). The rapid detection of bacteria and on-site analysis is necessary to reduce the public health risk posed by faecal contamination in recreational waters, thus making biosensors an important tool to achieving these goals. Most biosensors for the testing of microbial water quality that were developed to date utilise the established biochemical and genetic markers described above. The advantage of most biosensors is that they can detect these markers directly in the water sample without the need of a cultivation step. This makes the test more rapid, independent of microbiology laboratory facilities and the results include both viable and VBNC cells. This section will review the different analytical and technological approaches in biosensor research and discuss their role in the analysis of bacterial contamination in environmental waters.

Optical Biosensors

A common approach in biosensor technology is the use of specific antibodies for the detection of target analytes in a sample. Immobilised antibodies against specific bacterial surface molecules can allow for quantification of bacteria in a sample with high specificity and sensitivity. In the late 1990's immunosensors for bacterial pathogens were described based on a range of different detection systems, including surface plasmon resonance (Fratamico, 1998) and potentiometry (Gehring, 1998). The group by Lim at the University of South Florida first described an optical biosensor for the detection of *E. coli* O157:H7 in meat (DeMarco, 1999). This was based on fluorescent labelled polyclonal antibodies in a sandwich ELISA format and a fibre optic detection system. The sensor was able to detect 3-30 cfu/ml in 20 min. They developed further biosensors on the same platform for the detection of other food-borne pathogens and toxins. In 2002 they reported a biosensor capable of detection of *E. faecalis* in untreated sewage, with a sensitivity of 2×10^5 cfu/ml (Trindade, 2002). Immunosensors are limited to the detection of specific strains of bacteria, which makes them useful for quantifying pathogenic bacteria but less so for more diverse indicator organisms. In order to monitor recreational water contamination, lower sensitivity is required in order to achieve values that allow reporting compliance with water quality standards. By combining an ultrafiltration and centrifugation step to concentrate the sample with a biosensor, it was possible to reliably detect *Enterococci* in recreational water samples at the US EPA regulatory standard level of 105 cfu/ml (Leskinen, 2008). The same group have developed a portable device to concentrate water samples which can increase sensitivity and at the same time reduces variability as larger volume of water can be sampled (Leskinen, 2009; Leskinen, 2012).

The use of portable and sensitive fluorescent readers allows for rapid testing of water samples without the need for concentration or other pre-treatment. A study based on fluorescent detection of β-D-glucuronidase activity has shown that *E. coli* could be reliably detected in river water as low as 7 cfu/ml (Wildeboer, 2010). A field kit for Coliforms testing is available by the Norwegian company Colifast®, which provides results in 15 min to 2 h. The technology is based on an advanced growth medium combined with fluorescent β-D-galactosidase substrate and can detect as low as 1 cell per sample volume (http://www.colifast.no/products/colifast-field-kit/).

Electrochemical Biosensors

Early biosensors largely employed electrochemical signal transducers to quantify the biological or biochemical marker. Mittelmann (2002) describe an amperometric biosensor that can quantify β-D-galactosidase activity corresponding to 1000 cfu/ml generating results in 75 min. The system detected the chemical oxidation of the substrate, thus employing the same principle on which the first glucose biosensors were developed. The system proved however to be not sensitive enough to detect β-D-glucuronidase. A more sensitive sensor for the quantification of β-D-galactosidase that is capable of detecting 1 cfu/100 ml in an assay time of 15 min was described by Wutor (2007). Their technique is based on voltammetric detection of the formation of chlorophenol red using a copper phthalocyanine modified glassy carbon working electrode. The detection of phenol by an amperometric biosensor that incorporated tyrosinase as the bioelement and carbon nanotubes coated with magnetic nanoparticles achieved a detection limit of 10 cfu/ml (Cheng, 2009). A detection limit of 7.5 *E. coli*/ml was achieved with an immunosensor using potentiometry as signalling element (Ercole, 2002). This was based on a sandwich ELISA principle, with the secondary antibody labelled with urease and the pH increase measured by the electrode.

An alternative to the enzyme based detection is the incorporation of specific antibodies into a biosensor. In this alternative to the above described electrochemical or optical detection, impedance can be used to detect the binding of the bacterial antigen to the immobilised sensor bound antibodies. Such a system does not require labelled antibodies or chemical signal amplification and has been developed for *E. coli* O157:H7 with a LOD of 100 cfu/ml (Chowdhury, 2012). Another sensing element using ultrasound has been shown to detect 1.6×10^4 cfu/ml *E. coli* O157:H7 in 20 min, demonstrating another alternative signalling element that can be incorporated into a bacterial immunosensor (Bavli, 2014). However, as noted by the authors, sensitivity needs to be improved in order to be able to detect bacterial levels relevant for environmental water quality.

Aptamers are oligonucleotides, DNA or RNA, that specifically bind to molecules including lipopolysaccharides and outer membrane proteins. This is a new field of research and only a small number of aptamers against *E. coli* have been described thus far. Queiros (2013) have developed an impedance biosensor with DNA aptamers against *E. coli* outer membrance proteins and demonstrated that such a system has the potential for a rapid and sensitive detection without the need for sample extraction or labelling.

The combination of molecular analysis and biosensor technology holds many promises as it combines benefits discussed above. DNA biosensors have recently found their use in medical applications, e.g., for Malaria or HIV testing, and it has also been demonstrated that these can be used to detect bacteria in a sample. The advantage of DNA biosensors is the direct detection of the target DNA extracted from the sample, thus cutting out the amplification step. With a simplified DNA extraction protocol, testing can be done *in situ* and rapidly. The detection of *E. coli* at a range of 2-20 cfu/ml seawater with a DNA biosensor with two alternative detection systems, optical and electrochemical has been demonstrated (Paniel, 2013).

PHYSICOCHEMICAL METHODS

Flow cytometry is a rapid and sensitive method to quantitatively detect cells and specify them as well as determine viability. It is widely used in haematology and cell biology but also increasingly in microbiology. The aforementioned capacities are attractive for Environmental analyses as the presence of VBNC cells is a factor of uncertainty for most quantitative methods. The use of flow cytometry for the characterisation of bacterial cells was first described by van Dilla (1983) and 15 years later it was shown that the technology also has potential to assess bacterial viability (Nebe-von-Caron, 1998). Further research included the optimisation of fluorescent dyes and detection systems for different organisms. It was shown that *E. coli* O157:H7 can be detected as low as 100 cells/ml and that the technology is capable of detecting viable and VBNC cells in a complex sample (Khan, 2010). However, the technical requirement, required cell concentrations and reaction with relatively expensive fluorescent dyes currently means that flow cytometry can't be regarded as an alternative for routine monitoring of environmental waters but as an additional tool for specific problems. Further research and development might however result in flow cytometry becoming the method to achieve the most definite quantification of viable and non-viable cell numbers. The advances in flow cytometry and other methods in detecting and assessing the viability of bacterial endospores were reviewed by Mohapatra (2013).

The spectroscopic characterisation of chemical fingerprints is an alternative approach to detect and characterise bacteria in a sample. Even though some pioneering studies were published around 30 years ago, the more recent advances in technologies and computational data analysis have brought these methods to the forefront of Environmental pollution monitoring. UV Raman Spectroscopy uses pulsed light of excitation wavelengths around 220-260 nm and the resulting spectra correspond to nucleic acid and protein content of bacterial cells in a sample. These spectra can be successfully related to different bacterial species (Dalterio, 1987; Wu, 2001) and in combination with genetic database analysis even closely related species could be distinguished (Lopez-Diez, 2004). In-situ synthesis of Ag nanoparticles on the cell wall of bacteria has shown to enable sensitive and rapid detection of bacteria by surface-enhanced Raman scattering with a limit of detection of 250 cells/mL and an assay time of 10 min (Zhou, 2014). The advances in Raman spectroscopy make it an alternative detection method for biosensors. An immunoassay for *E. coli* was recently described to achieve quantitative detection as low as 4.3×10^3 cells/ml by antibody capture of the bacteria to the chip followed by SERS mapping (Knauer, 2012).

Proteomics methods have found many applications in microbiology and environmental studies. The method of choice is MALDI-TOF (matrix assisted laser desorption ionisation – time-of-flight) mass spectrometry as it can analyse whole cell samples, lysis or protein extraction are not required. Databases have been growing steadily allowing for classification and speciation of bacteria in a sample. New generation MALDI analysers like the Bruker Biotyper detect high abundance proteins and have an increased sensitivity, which can make concentration of the sample redundant. Together with bioinformatics tools contained in the system, microbe identification can be achieved in a few minutes (https://www.bruker.com/products/mass-spectrometry-and-separations/maldi-biotyper/overview.html). Proteomics has made and will continue to make important contributions to research and analysis of clinical and environmental samples, e.g., for the identification of pathogens, non-culturable organisms or antibiotic resistance, but is not suitable for routine monitoring of surface water quality.

CONCLUSION

The analysis of environmental waters for bacterial contamination has two main objectives, the quantification of indicator organisms in order to detect faecal pollution and the testing for specific pathogens that pose a risk to human health. Whereas methods have been refined and standardised and tolerable levels are more aligned globally, the analysis is largely still based on scientific concepts that have been known for many decades (Table 1). Largely, enumeration of intestinal *Enterococci* and *E. coli* by culture and detection of β-D-galactosidase and β-D-glucuronidase activity is the basis of routine analysis. Diagnostic enzymes for major pathogens that are associated with waterborne diseases are utilised in culture based biochemical tests. Bacterial identification and quantification by molecular methods has become the gold standard in many areas. For environmental samples there are however some limitations, notably the need to concentrate the sample or amplify the DNA and the highly specific target sequence that could result in an underestimation of the actual level of indicator organism. The high specificity is however a benefit when testing for individual pathogens and the fact that all cells, not just the culturable fraction, are detected is another advantage. A wide range of biosensor systems have been developed based on different biological sensing and physical signalling elements. Several show promising sensitivity as well as simplicity for direct and *in situ* analysis of environmental waters but it is apparent from the literature that more research is needed before any device will be able to complement or even replace the current standard methods for routine monitoring of water quality.

REFERENCES

Abdel Zaher AM, Wright ME, Ortega C et al. (2011). Daily measures of microbes and human health at a non-point marine beach. *J Water Health* 9: 443-457.

Adcock PW and Saint CP (2001). Rapid confirmation of *Clostridium perfringens* by using chromogenic and fluorogenic substrates. *Appl. Environ Micro.* 67: 4382-4384.

Alexander SK and Strete, D (2001). *Microbiology A photographic atlas for the laboratory*. Addison Wesley Longman, Inc USA.

Amirat L, Wildeboer D, Abuknesha RA and Price RG (2012). *Escherichia coli* contamination of the river Thames in different seasons and weather conditions. *Water and Environment J*, 26: 482-489.

Arnold BF, Schiff KC, Griffith JF et al. (2013). Swimmer illness associated with marine water exposure and water quality indicators: Impact of widely used assumptions. *Epidemiology* 24:845-853.

Atlas RM (2010). Handbook of Microbiological Media, 4th Ed., *CRC Press.*

AUS Gov NHMRC (2008). Guidelines for Managing Risks in Recreational Water. Australian Government National Health and Medical Research Council, Canberra.

Bavaro MF (2009). *Escherichia coli* 0157: What every internist and gastroenterologist should know. *Curr Gastroeneterology Rep* 11: 301-306.

Bavli D, Emanuel N and Barenholz Y (2014). Real-time monitoring of E. coli O157 and Salmonella enterica serovar Typhimurium in water using ultrasound and latex-based immunoassay. *Anal Methods* (2):395.

Beneduce L, Spano G and Massa E (2003). *Escherichia* coli 0157:H7 general characteristics. Isolation and identification techniques. *Ann of Micro* 53: 511-52.

Boehm A and Sassoubre LM (2014). *Enterococci* as indicators of environmental contamination. In: Commensals to leading causes of drug resistant infection. Gilmore MS, Clewell DB, Ike Y (eds), Boston, *Mass Eye and Ear Infirmary.*

Bottari B, Ercolini D, Gatti M and Neviani E (2006). Application of FISH technology for microbiological analysis: current state and prospects. *Appl Microbial Biotechnol* 73: 485-494.

Brettar I and Hofle MG (2008). Molecular assessment of bacterial pathogens – a contribution to drinking water safety. *Curr Opin Biotechnology* 19: 274-280.

Browne NK, Huang Z, Dockrell M, Hashmi P and Price RG (2010). Evaluation of new chromogenic substrates for the detection of coliforms. *J Appl Micro* 108: 1823-1838.

Call DR, Borucki MK and Loge JL (2003). Detection of bacterial pathogens in environmental samples using DNA arrays, *J Micro Meth* 53: 235-243.

Cassar R and Cuschieri P (2003) Comparison of Salmonella Chromogenic Medium with DCLS Agar for Isolation of Salmonella Species from Stool Specimens. *J Clin Micro* 41: 3229–3232.

Cenciarini-Borde C, Coutois S and La Scola B (2009). Nucleic markers for bacterial detection using molecular tools. *Future Micro* 4: 45-64.

Cheng Y, Liu Y, Huang J, Li K, Xian, Y, Zhang W and Jin, L. (2009). Amperometric tyrosinase biosensor based on Fe_3O_4 nanoparticles-coated carbon nanotubes nanocomposite for rapid detection of coliforms. *Electrochimica Acta* 54: 2588-2594.

Chowdhury AD, De A, Chaudhuri CR, Bandyopadhyay K and Sen P. (2012). Label free polyaniline based impedimetric biosensor for detection of E. coli O157:H7 Bacteria. *Sensors Actuators B: Chem* 171–172(0): 916-923.

Colford JM, Schiff KC, Griffith JF et al., (2007) Using rapid indicators for *Enteroccoccus* to assess the risk of illness after exposure to urban runoff contaminated marine water. *Water Res* 46: 2176-2186.

CONAMA (2000) Resolution CONAMA no 274. Conselho Nacional do Meio Ambiente, Ministerio Do Meio Ambiente, Governo Federal Brazil, Brasilia. Available at: www.mma.gov.br/port/conama/res/res00/res 27400.html

Cooke VM, Miles RJ, Price RG and Richardson AC (1999). A Novel Chromogenic Ester Agar Medium for Detection of Salmonellae. *Appl and Environ Micro* 65: 807–812.

Cooke VM, Miles RJ, Price RG, Midgley G, Khamri W and Richardson AC (2002). New Chromogenic Agar Medium for the Identification of *Candida* spp. *Appl and Environ Micro* 68: 622–3627.

Cordero L, Norat J, Mattei H and Nazario C (2012). Seasonal variations in the risk of gastrointestinal illness on a tropical recreation beach. *J Water Health* 10: 579-593.

Dalterio RA, Nelson WH, Britt D and Sperry JF (1987). An Ultraviolet (242 nm Excitation) Resonance Raman Study of Live Bacteria and Bacterial Components. *Appl Spectrosc* 41(3): 417-422.

DeMarco D, Saaski E, McCrae D and Lim D (1999). Rapid detection of *Escherichia coli* O157: H7 in ground beef using a fiber-optic biosensor. *J Food Prot* 62: 711-716.

Dong J, Olano JP, McBride JW and Walker DH (2008). Emerging pathogens: challenges and successes of molecular diagnostics (2008). *J Mol Diagn* 10: 185-197.

Dorevitch S, Pratap P, Wroblewski et al., (2012). Health risks of limited-contact water recreation. *Environ Health Perspect* 120: 192-197.

Edwards C, Diaper J and Porter J (1996). Flow cytometry for the targeted analysis of the structure and function of microbial populations. In: Pickup RW & Saunders JR (ed. Molecular approaches to environmental microbiology. *Hemel Hempstead, Ellis Horwood Ltd.*

Ercole A, Del Gallo M, Pantalone M, Santucci S, Mosiello L, Laconi C and Lepidi A (2002). A biosensor for *Escherichia coli* based on potentiometric alternating biosensing (PAB) transducer. *Sensors and Actuators B* 83: 48-52.

EU (2006). Directive 2006/7/EC (2006) of the European Parliament and of the council of 15 February 2006 concerning the management of bathing water quality. The European Parliament and the Council of the European Union: *Official Journal of the European Union* L64/37.

Fewtrell L and Kay D (2015). Recreational water and infection: A review of recent findings. *Curr Environ Health Rpt* 2: 85-94.

Fleisher JM, Fleming LE, Solo-Gariele HM et al., (2010). The BEACHES Study: health effects and exposures from non-point source microbial contaminants in subtropical recreational marine waters. *Int J Epidemiol* 39: 1291-1298.

Fratamico PM, Strobaugh T, Medina M and Gehring A (1998). Detection of *Escherichia coli* O157: H7 using a surface plasmon resonance biosensor. *Biotechnol Tech* 12: 571-576.

Gehring AG, Patterson DL and Tu S (1998). Use of a Light-Addressable Potentiometric Sensor for the Detection of *Escherichia coli* O157:H7. *Anal Biochem* 258: 293-298.

Girones T, Ferrus MA, Alonso JL, Rodriguez-Manzano J, Calgua B, Correa Ade A, Hundesa A, Carratala A and Bofill-Mas S (2010). Molecular detection of pathogens in water – the pros and cons of molecular techniques. *Water Res* 4: 325-4339.

Godoy P, Nuín C, Alsedà M, Llovet T, Mazana R and Domínguez A (2006). Waterborne outbreak of gastroenteritis caused by *Norovirus* transmitted through drinking water. *Revista Clínica Española* 206(9): 435–437.

Gomi R, Matsuda T, Matsui Y and Yoneda M (2014). Fecal source tracking in water by next-generation sequencing technologies using host-specific *Escherichia coli* genetic markers. *Environ Sci Technol* 48(16): 9616-9623.

Hall V (2013). Epidemiological investigation of an outbreak of gastroenteritis following a mass participation swim in the River Thames in 2012. Public Health England.

Heijnen L and Medema G (2006). Quantitative detection of *E. coli* 0157 and other shiga toxin producing *E. coli* in water samples using culture method combined with real-time PCR. *J Water Health* 4: 487-498.

Ibrahim EH, Sherman G, Ward S, Fraser VJ and Kollef MH (2001). The influence of inadequate antimicrobial treatment of bloodstream infections on patient outcomes in the ICU setting. *Chest* 118: 146-55.

ISO (2015). International Standards Organisation: ISO/TC 147/SC 4 - Microbiological methods. Available at: http://www.iso.org/iso/home/store/catalogue_tc/catalogue_tc_browse.htm?commid=52944&published=on

Juhna T, Birzniece D, Larsson S, Zulenkova D, Shapiro A, Azevedo NF, Menard-Szczebara F, Castagnel S, Feliers C and Keevil CW (2007). Detection of *Escherichia coli* in biofilms from pipe samples and coupons in drinking water distribution networks. *Appl Environ Micro* 73: 7456-7464.

Khan MMT, Pyle BH and Camper AK (2010). Specific and Rapid Enumeration of Viable but Nonculturable and Viable-Culturable Gram-Negative Bacteria by Using Flow Cytometry. *Appl Environ Micro* 76(15): 5088-5096.

Knauer M, Ivleva N, Niessner R and Haisch C (2012). A flow-through microarray cell for the online SERS detection of antibody-captured *E. coli* bacteria. *Anal Bioanal Chem* 402(8): 2663-2667.

Kolarević S, Knežević-Vukčević J, Paunović M, Gačić Z and Vuković-Gačić B (2011). Assessment of the Microbiological Quality of the River Tisa in Serbia. *Water Res and Management* 1: 57-61.

Leskinen SD and Lim DV (2008). Rapid Ultrafiltration Concentration and Biosensor Detection of *Enterococci* from Large Volumes of Florida Recreational Water. *Appl Environ Micro* 74: 4792-4798.

Leskinen SD, Harwood VJ and Lim DV (2009). Rapid dead-end ultrafiltration concentration and biosensor detection of *enterococci* from beach waters of Southern California. *J Water Health* 7: 674-684.

Leskinen SD, Kearns EA, Jones WL, Miller RS, Bevitas CR, Kingsley MT, Brigmon RL and Lim DV (2012). Automated dead-end ultrafiltration of large volume water samples to enable detection of low-level targets and reduce sample variability. *J Appl Micro* 113: 351-360.

Lodder WJ and de Roda Husman AM (2005). Presence of Noroviruses and Other Enteric Viruses in Sewage and surface waters in the Netherlands. *Appl Environ Micro* 71: 1453-1461.

Lopez-Diez EJ and Goodacre R (2004). Characterization of Microorganisms Using UV Resonance Raman Spectroscopy and Chemometrics. *Anal Chem* 76: 585-591.

Manafi M, Kneifel W and Bascombe S (1991). Fluorogenic and chromogenic substrates used in bacterial diagnostics. *Microbiol Rev* 55: 2011-234.

McLellan SL and Eren AM (2014). Discovering new indicators of fecal pollution. *Trends Microbiol* 22(12): 697-706.

McPeters GA, Cameron SC and Le Chevallier MW (1982). Influence of diluents, media and membrane filters on detection of injured waterborne coliform bacteria. *Appl Environ Micro* 43: 97-103.

Mittelmann AS, Ron EZ and Rishpon J (2002). Amperometric Quantification of Total Coliforms and Specific Detection of *Escherichia coli*. *Anal Chem* 74: 903-907.

Mohapatra BR and La Duc MT (2013). Detecting the dormant: a review of recent advances in molecular techniques for assessing the viability of bacterial endospores. *Appl Microbiol Biotechnol* 97(18): 7963-7975.

Nebe-von-Caron G, Stephens P and Badley R (1998). Assessment of bacterial viability status by flow cytometry and single cell sorting. *J Appl Microbiol* 84(6): 988-998.

Nebe-von-Caron G, Stephens PJ, Hewitt CJ, Powell JR and Badley RA (2000). Analysis of bacterial function by multi-colour fluorescence flow cytometry and single cell sorting, *J Micro Meth* 42: 97-114.

Noble RT and Weisberg SB (2005). A review of the technologies for rapid detection of bacteria in recreational waters. *J Water Health* 3: 381-392.

OECD (2004). Environment, Health and Safety Publications, Series on Harmonisation of Regulatory Oversight in Biotechnology No 30; Guidance on methods for detection of microorganisms introduced into the environment: *Bacteria.*

Pall E, Niculae M, Kiss T, Sandru CD and Spinu M (2013). Human impact on the microbiological water quality of the rivers. *J Med Micro* 62: 1635–1640.

Paniel, N. and Baudart, J. (2013). Colorimetric and electrochemical genosensors for the detection of *Escherichia coli* DNA without amplification in seawater. *Talanta* 115: 133-142.

Perez JM, Cavalli P, Rourde C, Renac R. Gille Y. and Freydiere AM (2003). Comparison of Four Chromogenic Media and Hektoen Agar for Detection and Presumptive Identification of *Salmonella* Strains in Human Stool. *J Clin Microbiol* 41: 1130-1134.

Pernthaler JFO, Glockner W, Schönhuber W and Amann R (2007). Fluorescence *in situ* hybridization In: J. Paul (ed.), Methods in Microbiology: Marine Microbiology, Vol. 30. *Academic Press Ltd,* London.

Perry JD, Rennison C, Butterworth LA, Hopley AL, Gould FK (2003). Evaluation of *S. aureus* ID, a new chromogenic agar medium for detection of *Staphylococcus aureus*. *J Clin Microbiol* 41: 5695-8.

Queirós RB, de-los-Santos-Álvarez N, Noronha JP and Sales MGF (2013). A label-free DNA aptamer-based impedance biosensor for the detection of E. coli outer membrane proteins. *Sensors Actuators B: Chem* 181(0): 766-772.

Ryzinska-Paier G, Lendenfeld T, Correa K, Stadler P and Blaschke AP, Mach RL, Stadler H, Kirschner AKT and Farnleitner AH (2014). A sensitive and robust method for automated on-line monitoring of enzymatic activities in water and water resources. *Water Science and Technolog*y 69: 1349-1358.

SA EA (2012). South African Water Quality Guidelines for Coastal Marine Waters. Volume 2: Guidelines for Recreational use. Department: Environmental Affairs, Republic of South Africa.

Samra Z, Heifets M, Talmore J, Bain E and Bahar J (1998). Evaluation of Use of a New Chromogenic Agar in Detection of Urinary Tract Pathogens. *J Clin Micro* 36: 990–994.

Scholes L, Faulkner H, Downward S and Tapsell T (2008). Urban Rivers as Pollutant Sinks and Sources: a Public Health Concern for Recreational River Users? *Water, Air, Soil Pollution: Focus* 8(5-6): 543-553.

Sidhu JPS and Toze SG (2009). Human pathogens and their indicators in biosolids: a literature review. *Environ Int* 35: 187-201.

Skariyachan S, Mahajanakatti AB, Grandhi NJ, Prasanna A, Sen B, Sharma N, Vasist KS and Narayanappa S. (2015). Environmental monitoring of bacterial contamination and antibiotic resistance patterns of the faecal coliforms isolated from Cauvery River, a major drinking water source in Karnataka, India. *Environ Monit Assess* 187: 279.

Song KY, Hyeon JY, Shin HC, Park CK, Choi IS, Seo KH (2008). Evaluation of a chromogenic medium supplemented with glucose for detecting *Enterobacter sakazakii*. *J Micro Biotech* 18: 579-84.

Tallon P, Magajna B, Lofranco C and Leung KT (2005). Microbial indicators of faecal contamination in water. *Water, Air and Soil Pollution* 166: 139-160.

Tornevi A, Bergstedt O and Forsberg B (2014). Precipitation Effects on Microbial Pollution in a River: Lag Structures and Seasonal Effect Modification. P*losOne* 9: e98546.

Trindade T, Kramer MF and Lim DV (2002). Antibody based detection of fecal *enterococci* utilizing an evanescent wave fibre optic biosensor. *Abstr Gen Meeting Am Soc for Micro* 102: 437.

Tryland I, Surman S and Berg JD (2002). Monitoring Faecal contamination of the Thames Estuary using semi-automated early warning system. *Water Sci Technol* 46: 25-31.

UK Environment Agency (2010). The River Basin Districts Typology, Standards and Groundwater threshold values (Water Framework Directive).

UK Environment Agency (2015). Bathing water quality. http://apps.environment-agency.gov.uk/wiyby/138153.aspx

US EPA (2002a). Method 1600: *Enterococci* in water by membrane filtration using membrane-*enterococcus* indoxyl-β-D-glucoside agar (mEI). Publication EPA-821- R-02-022. USEPA Office of Water, Office of Science and Technology, USEPA, Washington, DC.

US EPA (2002b). Method 1603: *Escherichia coli (E. coli)* in water by membrane filtration using membrane-thermotolerant *Escherichia coli* agar (Modified mTEC). Publication EPA-821- R-02-023. USEPA Office of Water, Office of Science and Technology, USEPA, Washington, DC.

US EPA (2012). *Recreational water quality criteria.* Office of Water, 820-F-12-058.

Van Dilla MA, Langlois RG, Pinkel D, Yajko D and Hadley WK (1983). Bacterial characterization by flow cytometry. *Science* 220(4597): 620-622.

Victor R, Lachica F, Weiss KF and Deibel RH (1969). Relationships among coagulase, enterotoxin, and heat-stable deoxyribonuclease production by *Staphylococcus aureus*. *Appl Microbiol.* 18:126-127.

Westrell T, Ciampa A, Boelaert F, Helwigh B, Korsgaard H, Chriel M, Ammon A AND Makela P (2009). Zoonotic infections in Europe in 2007: a summary of ETSA-ECDC annual report. *Euro Surveill* 14 (3): pii 19100.

WHO (2003). Chapter 4: Faecal pollution and water quality. In: Guidelines for safe recreational water environments. Volume 1: Coastal and fresh waters. World Health Organisation, available from: http://www.who.int/water_sanitation_health/bathing/srwe1/en/

Wildeboer D, Amirat L, Price RG and Abuknesha RA (2010). Rapid detection of *Escherichia coli* in water using a hand-held fluorescence detector. *Water Res* 44(8): 2621-2628.

Wu Q, Hamilton T, Nelson WH, Elliott S, Sperry JF and Wu M (2001). UV Raman spectral intensities of *E. coli* and other bacteria excited at 228.9, 244.0, and 248.2 nm. *Anal Chem* 73(14): 3432-3440.

Wutor VC, Togo CA, Limson JL and Pletschke BI (2007). A novel biosensor for the detection and monitoring of β-D-galactosidase of faecal origin in water. *Enzyme Micro Technol* 40: 1512-1517.

Xu S and Mutharasan R (2011). The Coming Together of the Sciences: Biosensors for the Detection of Waterborne Pathogens Using Antibodies and Gene-based Recognition Chemistries. In: *Environmental Microbiology – Current Technology and Water Applications* by K. Sen and N.J. Ashbolt (Eds). Caister Academic Press, Norfolk, UK.

Zhou H, Yang D, Ivleva NP, Mircescu NE, Niessner R and Haisch C (2014). SERS Detection of Bacteria in Water by in Situ Coating with Ag Nanoparticles. *Anal Chem* 86(3): 1525-1533.

Zuckermann U, Hart I and Armon R (2008). Field evaluation of ground, raw and treated water and comparison with standard membrane filtration methods. *Water, Air, Soil Pollut* 188: 3-8.

In: Coliforms
Editor: Geoffrey McCoy
ISBN: 978-1-63483-542-8

Chapter 3

TESTING OF COLIFORMIC BACTERIA FOR WATER DEPARTMENTS USING AUTOMATED SAMPLING AND ANALYSIS

Elias Hakalehto[1,2], Anneli Heitto[2], Osmo Hyyrynen[3], Juha Jokelainen[4], Ari Jääskeläinen[5] and Lauri Heitto[6]

[1]Department of Environmental Sciences,
University of Eastern Finland, Kuopio, Finland
[2]Finnoflag Oy, Kuopio, Finland
[3]Veseco Oy, Luumäki, Finland
[4]Samplion Oy, Kuopio, Finland
[5]Savonia University of Applied Sciences, Kuopio, Finland
[6]Environmental Research of Savo-Karjala, Kuopio, Finland

ABSTRACT

In municipal or household water production and in its monitoring, there is a need to protect the consumers from dangerous microbial impacts. It is important to find out even the smallest concentrations of hazardous organisms or their toxins. In order to screen huge masses of water flow, enrichment technologies and indicator organisms are used according to international guidelines. Also, automated sampling and analysis—as well as data transfer—are warranted. The PMEU technologies have been developed for enhanced microbial cultivation and detection. The Coliline PMEU version is developed for automated water monitoring.

1. Introduction

The PMEU (Portable Microbe Enrichment Unit) enhances the growth of microbes. Special sampling syringes that contain samples are placed into the PMEU and a sterile gas mixture is led into the syringes. The specific cultivation and carrier gases normally play an important role in producing the enrichment result. An optimal growth environment for the microbes is created in the syringe by using microbe-specific culture media and by changing the mixture and dosage of the gas, as well as the sample temperature. Several different samples can be enriched simultaneously and the PMEU enables the start of the enrichment phase directly after taking the samples. This can be done also outside of laboratories in field conditions.

The development of the PMEU represents the ideas of how modern microbiological research work could create new innovative products and methods to be used in microbe diagnostics as a sector of biotechnology development. Microbial detection tests are carried out all around the world. Application areas cover a wide range from preventive cleanliness of products and production environments to rapid detection of bacteria that cause illnesses; from environmental control to security applications. Microbe diagnostics allow us to understand our living conditions better and to cope with the challenges or threats posed to us.

Microbe diagnostics is also a growing business area. A private market study estimates the world's potential for microbiological tests carried out in the food industry alone at around 650 million tests annually with a market value at almost USD 1 billion. Apart from food industry, microbe diagnostics is used in at least ten other industries including clinical and environmental sectors and agriculture. The testing of raw water is crucial for hygiene maintenance. The early warning about microbial contaminations is crucial in safe-guarding general public health. Early characterization of antibiotic resistance markers is also increasingly important (Hakalehto, 2015a; Hakalehto et al., 2015a, b, c, d).

The PMEU solves many of the problems related to conventional microbial diagnostics. It helps to understand microbe detection as a process, where each step is important (Figure 1). In addition it is compatible with any conventional final standard analysis, method or equipment in use today. The PMEU may be used in field conditions as a miniature mobile lab or in an actual laboratory as a cultivator and detector device for boosting microbial growth. Wherever used, it brings efficiency to the microbe detection process, allowing the cultivation of both aerobic and anaerobic microbes take place in the most optimal growth environment.

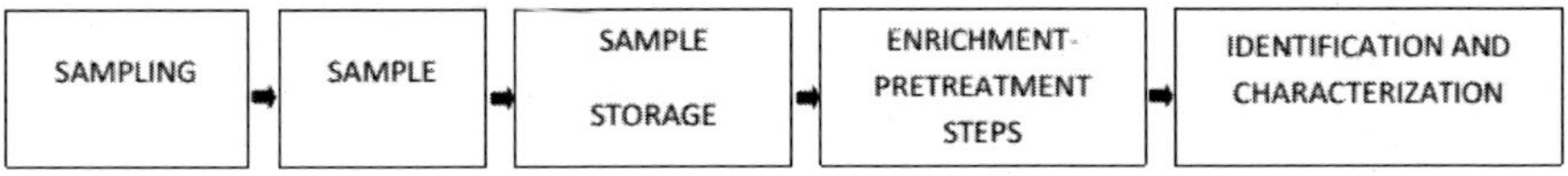

Figure 1. Microbial detection is a sequential process consisting of several equally important steps.

Apart from microbiological innovations the PMEU integrates also the advances of electronics and mobile communications technologies to make it the ideal device for efficient microbe diagnostics.

Current microbiological problems that existing conventional microbiology cannot solve, relate to the following questions:

1. How to ensure the survival of microbes during and after sampling and transport?
2. How to make microbes grow as optimally as possible?
3. How to detect microbes from initially small concentrations?

As an example, *Salmonella* detection with traditional methods from sampling through to final analysis takes approximately 72 hours. With the PMEU method the time is diminished to about 5-15 hours.

PMEU (Portable Microbe Enrichment Technology) has been developed for starting microbiological sample processing:

- at sampling sites
- on the way to laboratory
- as an integrated method for detection and in cohesion with any other laboratory analysis methods
- with remote-control facilities
- for simulating microbiological metabolism and interactions in natural conditions
- in developing industrial processes
- as the fastest method to observe microbial growth

- for increased speed, reliability and accuracy
- in scientific microbiological research
- for environmental and food monitoring
- for increased security
- for hospital hygiene
- in instant patient care
- for personalized medicine solutions

In the field use of the PMEU, enrichment starts immediately after sampling. The specimen collection may take place automatically, and it can be remote-controlled. This device is applicable outdoors, in industrial plants, by the production lines, in control rooms and situations, and in laboratories. The cultivation of samples can be started during their transportation and storage if necessary. Consequently, the samples stay in a representative condition. The method is applicable for both aerobic and anaerobic microorganisms. The mixing of gases and nutrients, incubation temperature and flow rate of air/gas can be controlled.

On the basis of the above-mentioned arrangements, faster and more accurate results can be obtained. The PMEU method shortens (depending on the microbial species) the detection periods for several hours or even days (Hakalehto, 2013; Hakalehto et al., 2014). Minor sub-populations derived from mixed populations can also be rapidly detected and identified (Hakalehto et al., 2010). The PMEU makes the enhanced detection of environmentally stressed microbes possible (Pesola et al., 2009; Pesola and Hakalehto, 2011; Hakalehto et al., 2011). It is possible to enrich antigens and other biomolecules as well.

2. History of Hygiene Indication and the Present Situation

Demand for the uniformity of microbiological methods started from water hygiene analysis. In year 1895, the American Public Health Association recognized the need to diminish or abolish large laboratory-to-laboratory variation in diagnostic methods, and appointed a committee to draft uniform procedures to some chemical determinations (Doyle et al., 2001). In 1905, also microbiological methods were included in a committee meeting regarding water analysis, producing a manual, Standard Methods of Water Analysis.

Interestingly, this manual was published only 13 years after the description of one remarkable fecal bacterium *Bacillus coli* (nowadays *Escherichia coli*) by T. Escherich. The concept of "fecal coliforms" was introduced in 1904 by C. Eijkman. The species *E. coli* and the group of thermotolerant fecal coliformic bacteria (bacteria resembling *E. coli*) are the basic hygienic indicators of water quality, especially with respect to contamination control. However, it is essential to recognize that the water and food qualities are reflected by the presence or absence of many other bacterial groups, besides the coliforms or the fecal strains (Hakalehto, 2015).

In the present times, WHO Guidelines for drinking-water quality - 4th ed. states the following about microbial water quality.

"For microbial water quality, verification is likely to be based on the analysis of faecal indicator microorganisms, with the organism of choice being *Escherichia coli* or, alternatively, thermotolerant coliforms (see sections 4.3.1, 7.4 and 11.6). Monitoring of specific pathogens may be included on very limited occasions to verify that an outbreak was waterborne or that a WSP has been effective. *Escherichia coli* provides conclusive evidence of recent faecal pollution and should not be present in drinking water.

Under certain circumstances, additional indicators, such as bacteriophages or bacterial spores, may be used.

However, water quality can vary rapidly, and all systems are at risk of occasional failure. For example, rainfall can greatly increase the levels of microbial contamination in source waters, and waterborne outbreaks often occur following rainfall. Results of analytical testing must be interpreted taking this into account.

The greatest risk to public health from microbes in water is associated with consumption of drinking-water that is contaminated with human and animal excreta, although other sources and routes of exposure may also be significant. Infectious diseases caused by pathogenic bacteria, viruses, protozoa and helminths are the most common and widespread health risk associated with drinking-water." (WHO, 2011).

3. Some Potentials of Finnoflag PMEU Technology

The basic principle of the PMEU is to make the processing of microbiological samples to happen immediately after the sampling, or as soon as possible. By this approach the vital information on the samples is obtained without the cells

a. losing their vitality
b. changing their characteristics
c. causing overgrowth of some species at the expense of the others

Correspondingly, the following goals are achieved:

a. optimal recovery of e.g., environmentally stressed microbial cells
b. optimal growth of all strains of interest
c. handy ways for carrying out the microbiological dilution series as well as the division of samples or cultures into further studies

On the basis of the above-mentioned benefits, the PMEU technology simplifies, intensifies, speeds up, and improves both the routine and research microbiology by:

a. throughput of analysis is multiplied
b. each sample is more carefully studied
c. further studies can be started instantaneously
d. all data is handled and transferred in real-time
e. laboratory safety is elevated

Also the surface related phenomena, such as biofilm formation, can be studied in the PMEU.

By using the ASCS (Automated Sample Collection System) remote control of sampling of liquids and gases can be achieved (Hakalehto, 2011; Hakalehto and Heitto, 2012). This makes it possible to establish systems and networks for automated microbiological monitoring in hospitals, industrial plants, and environment.

This kind of a system has been developed to the Coliline PMEU™ version (Figure 2). Several such units can be positioned into strategic locations along the water distribution network. They can be interlinked, or connected to an automated disinfection system which can be controlled by a computer according to the online PMEU data.

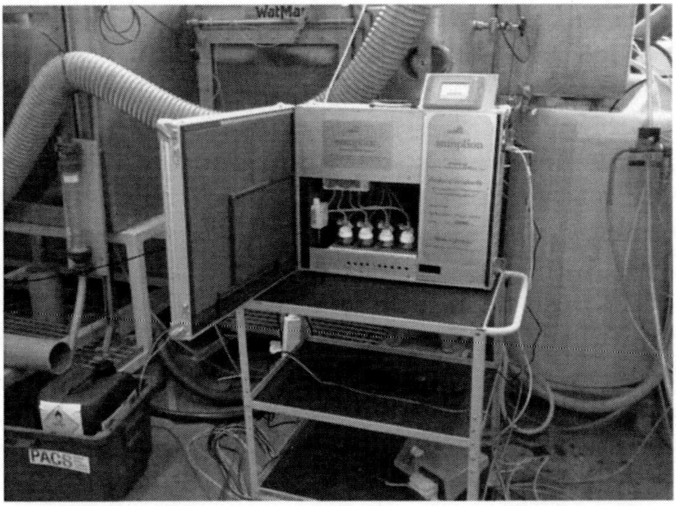

Figure 2. Remote-controlled Coliline^{TM} PMEU is automatically collecting samples from water production. It is globally represented by Berner Oy, Helsinki, Finland.

4. General Background

One area of a remarkable influence of the microbes is well seen: it is the pathogenic effect on man. Huge amounts of different kinds of pathogens are included in the infections and they are distributing via the waterways or food supply. And to detect them we need methods and tools. The detection itself is not a single event but it is a kind of process, it contains a few steps and the process is as effective as its weakest step. So we have to focus on each and every step very carefully. For example the importance of sampling is often neglected. We need to collect the specimens carefully considering all the circumstances and physical or chemical parameters etc. Then we need to transport the sample and that is the step when very often most of the microbes in the sample are being destroyed, they simply cannot sustain that transport step. This is one of the reasons, why we started developing a kind of instrument – Portable Microbe Enrichment Unit – for preparing the samples in a way that the smallest amount of information is lost and also remarkable amounts of time are saved. With the PMEU detection time is usually shortened at least by 50%, especially when boosting with appropriate gas flow is used. Very often we can say that we can shorten the detection from weeks to days and from days to hours.

We have done development work on the PMEU technology and its many variations. In different PMEU variants optical sensors, infrared or ultraviolet sensors and also the detection of the volatile compounds are used for the surveillance. Samples are collected into specific syringes and then the syringes are positioned into the equipment.

The PMEU method can be used for evaluating the results of disinfection (Hakalehto, 2010).

All various kinds of bacteria can be monitored by the PMEU and also yeasts, molds, any kinds of microbes and it is not only the question of detecting the bacteria but it is a matter of understanding how active they are. Metabolism is very important to be followed up in order to understand for example a phenomena like pathogenesis.

The PMEU equipment has been involved in an extensive nationwide project in Finland called the Polaris study. The equipment was tested in a couple of larger cities and some smaller towns and in all of them water departments have seen it functioning as an integral part of the standard hygiene monitoring or simultaneously with the Polaris tests. All participants of the project have been able to follow up the results and the enrichment real-time via the internet. Subsequently the growth is displayed as a standard growth curve.

The PMEU has been tested in several directions also abroad, including Burkina Faso in Central Africa where an existing problem was the detection of the pathogens that were beneath some stronger microbial growth, such as the coliforms. In this type of a situation it is necessary to develop methods beneath the usual hygiene indication (Hakalehto and Heitto, 2012). These methods could exploit the same sampling functions as applied to the monitoring of hygiene indicators. The pathogens transmit via the irrigation water and were potentially contaminating the vegetables and causing epidemics of *Yersinia* sp. or *Campylobacter* sp. (Hakalehto et al., 2014). This method, that offers a certain selectivity and enrichment of the pathogens, is needed. These pathogens were amongst the huge majority of other bacterial growth in pools to which all the waste waters of a city were directed. Without any treatment it was used for irrigation so it was obvious that the pathogens were being distributed.

These were some aspects of the detection but the detection is only the starting point after which we have to characterize our isolates in order to understand their actions. We need to comprehend how they function in the environment or in our bodies. In our intestines there are trillions of different bacterial cells in action all the time. What are they really doing and what's

If samples are to be obtained automatically, water that can originate from a water tube must circulate through the PMEU Coliline™. For each sample slot, a timer can be set between 0 and 300 hours. This way a sample can be taken once a day, for example.

In manual mode, samples can be taken from any water source, and they can be left to incubate inside the PMEU Coliline™. This mode also allows a tracking number and growth of potential coliforms and *E.coli*.

6.3. Observing Results

Once taking of samples has commenced, results can be tracked in real-time on the PMEU Coliline™ screen. Remote tracking and remote control of device is also possible, by using Ethernet connection, or wirelessly by adding a 3G comms package. The 3G comms package also makes alerts possible as emails and text messages. Results can be saved on a USB flash drive and thus examined afterwards on a computer.

6.4. Differences between PMEU Coliline™ and the Bubble-Flow Generating PMEU´s

The importance of the built-in bubble flow in the PMEU was verified in several studies with the PMEU (Hakalehto, 2013; Hakalehto et al., 2013). Also in the case of *Campylobacter* sp. the PMEU enrichment method together with the real-time PCR was found to be more effective a method for the detection of pathogens in water than the corresponding ISO standard method (Pitkänen et al., 2009).

In order to further exploit the use of PMEU technology for pathogen screening, we performed a field study in Burkina Faso, Western Africa, for identifying contaminating *Campylobacter* sp. and *Yersinia* sp. in municipal waste water used for irrigation and in the vegetables for sale in the local market in Ouagadougou (Hakalehto et al., 2014a). In this latter case, the microaerobic bubble flow was provided with a field capable method not dependent on the use of pressurized gas bottles.

Also the enhanced mycobacterial cultivation was made possible with the PMEU bubble flow (Hakalehto, 2013; Hakalehto et al., 2014b).

In addition to the microaerobic bacteria, or facultative anaerobes, the PMEU has been used for detecting strictly anaerobic pathogens, such as

Clostridium difficile (Hell et al., 2010), and aerobes, such as *Bacillus* sp. (Mentu et al., 2009). In any case, the formation of gas bubbles has been a key factor in speeding up the growth or enhancing the enrichment.

In case of *Salmonella* sp. it was proven that the anaerobic metabolism was equally effective as the aerobic one in maintaining cell growth (Hakalehto et al., 2007). However, the cell surface antigens can be expressed depending on the selected gas flow program and mode. In case of *Enterococcus* sp. in the factory waste water treatment pools, the PMEU turned out to favor somewhat different subpopulations than the traditional laboratory methods if oxygen was applied to the cultures in the gas flow (results not shown here). In general, it could be stated that a variation in the method of enrichment can cause qualitative variation in the results, not only quantitative alterations, in the composition of the microflora. This has to be taken into account in the source-tracking studies (Heitto et al., 2009).

The antibiotic resistance in the chyme populations was studied with the PMEU Spectrion® (Hakalehto, 2011) and PMEU Scentrion® (Laitiomäki et al., 2014).

7. The High Speed and Sensitivity in Bacterial Cultures: PMEU Scentrion ®

PMEU Scentrion® (Figure 5) has been used for pathogen detection in the hospital setting (Hakalehto et al., 2009; Pesola et al., 2012; Laitiomäki et al., 2014). It can be used also for food and water monitoring, when ultimate speed is required (Hakalehto et al., 2011; Hakalehto and Heitto, 2012). Some extra benefits obtained by the use of the PMEU Scentrion®.

- PMEU Scentrion® gives real-time analysis of growth by monitoring the emitted gases from the cultures
- PMEU Scentrion® can transmit data in the form of several simultaneous growth curves in real time, reflecting various aspects of the cultures, or different strains in a mixed culture.
- PMEU Scentrion® can be used for the detection of VOC compounds from awakening microbial cultures.

One of the most important things is to have the visualization of the bacterial growth on line from the screen to evaluate the time point to take sample(s) for other investigations. PMEU Scentrion® records a log-file from each individual sample according to the sampling time points. Typically, when 10 simultaneous samples are growing in the device, a new measurement is taken after 35-40 minutes from each sample. Thus it is possible to generate a growth-curve with a suitable computing program. According to the responses of measuring sensor cells to different metabolites of different bacteria it is possible to classify the microbes if the reference data is available. See Chapter 7 for the PMEU Scentrion® based results of chicken meat spoilage by *Salmonella* contaminants. PMEU Scentrion® was used for detecting a single *Salmonella* cell in tap water in 10 hours (Hakalehto et al., 2011). This method based on volatile detection can thus be implemented also into water hygiene monitoring.

Figure 5. A view on the PMEU Scentrion®. Photograph by Vicente Serra.

Conclusion

Emerging challenges in microbial diagnostics require:

1. Sampling improvements
2. Enhanced enrichment
3. Sped up recovery and detection of microbial cells in culture
4. Precise identification and data bank
5. Characterization elaboration
6. On-line monitoring and control system
7. Modern information transfer and management

The PMEU technologies have been developed in order to provide a tool and a platform to meet these challenges.

References

Doyle, M.P.; Beuchat, L.R.; Montville, T.J. (2001). Food Microbiology. Fundamentals and Frontiers. 2ed. Washington D.C., USA: ASM Press.

Hakalehto, E. (2010). Hygiene monitoring with the Portable Microbe Enrichment Unit (PMEU). *41st R3 -Nordic Symposium. Cleanroom technology, contamination control and cleaning. VTT Publications 266.* Espoo, Finland: VTT (State Research Centre of Finland). pp. 164-176.

Hakalehto, E. (2011). Simulation of enhanced growth and metabolism of intestinal *Escherichia coli* in the Portable Microbe Enrichment Unit (PMEU). In: Rogers MC, Peterson ND (eds.) E. coli infections: causes, treatment and prevention. New York, USA: Nova Science Publishers, pp.159-175.

Hakalehto, E. (2013). Enhanced mycobacterial diagnostics in liquid medium by microaerobic bubble flow in Portable Microbe Enrichment Unit1. *Pathophysiology* 20: 177–180.

Hakalehto, E. (2015). Mycobacterial Detection Views. In Hakalehto, E. (Ed.) Microbiological Clinical Hygiene, Nova Science Publishers Inc., N.Y., USA.

Hakalehto, E. and Heitto, L. (2012). Minute microbial levels detection in water samples by Portable Microbe Enrichment Unit Technology. *Environment and Natural Resources Research,* 2012; 2, 80-88.

Hakalehto, E.; Pesola, J.; Heitto, L.; Närvänen, A.; Heitto, A. (2007). Aerobic and anaerobic growth modes and expression of type 1 fimbriae in Salmonella, Pathophysiology 14(1); 61-69.

Hakalehto, E.; Pesola, J.; Heitto, A.; Deo, B. B.; Rissanen, K.; Sankilampi, U.; Humppi, T.; Paakkanen, H. (2009). Fast detection of bacterial growth by using Portable Microbe Enrichment Unit (PMEU) and ChemPro100i® gas sensor. *Pathophysiology, 16*, 57-62

Hakalehto, E.; Hell, M.; Bernhofer, C.; Heitto, A.; Pesola, J.; Humppi, T.; Paakkanen, H. (2010). Growth and gaseous emissions of pure and mixed small intestinal bacterial cultures: Effects of bile and vancomycin. *Pathophysiology, 17*, 45-53. http://dx.doi.org/10.1016/j.pathophys.2009.07.003.

Hakalehto, E.; Heitto, A.; Heitto L. (2011). Vesien Mikrobiaapinen ("A Microbiological ABC book of the waters") (in Finnish). Kuopio, Finland: Finnoflag Oy.

Hakalehto, E.; Heitto, A.; Heitto, L. (2013). Fast coliform detection in portable microbe enrichment unit (PMEU) with Colilert(®) medium and bubbling. *Pathophysiology*. 2013 Sep; 20(4):257-62. doi: 10.1016/j.pathophys.2013.05.001. Epub 2013 Jun 20.

Hakalehto, E.; Nyholm, O.; Bonkoungou, I.J.O.; Kagambega, A.; Rissanen, K.; Heitto, A.; Barro, N.; Haukka, K. (2014). Development of microbiological field methodology for water and food-chain hygiene analysis of *Campylobacter* spp. and *Yersinia* spp. in Burkina Faso, West Africa. *Pathophysiology*, 21: 219-229.

Hakalehto, E.; Heitto, A.; Heitto, L.; Rissanen, K.; Pesola, I.; Pesola, J. (2014b). Enhanced recovery, enrichment and detection of *Mycobacterium marinum* with the Portable Microbe Enrichment Unit (PMEU). *Pathophysiology*. 2014 Sep; 21(3):231-5. doi: 10.1016/j.pathophys.2014.07.005. Epub 2014 Aug 4.

Hakalehto, E.; Pesola, J.; Puhakainen, E. (2015a). Trends in Clinical Microbiology toward Point-of-Care Diagnostics. In Hakalehto, E. (Ed.) Microbiological Clinical Hygiene, Nova Science Publishers Inc., N.Y., USA.

Hakalehto, E.; Pesola, J.; Heitto, L. (2015b). Environmental Sampling and Sample Processing in Hospitals. In Hakalehto, E. (Ed.) Microbiological Clinical Hygiene. Nova Science Publishers Inc., N.Y., USA.

Hakalehto, E.; Hell, M.; Heitto, A. (2015c). Planning and Construction of Healthcare Facilities – a case evaluation on hygienic aspects. In:

Hakalehto, E., editor. Microbiological Clinical Hygiene. New York, NY, USA: Nova Science Publishers, Inc.; 2015; pp. 79-84.

Hakalehto, E.; Pesola, I.; Pesola, J.; Hell, M. (2015d). Antibiotic Resistance Essentials. In: Hakalehto, E., editor. Microbiological Clinical Hygiene. New York, NY, USA: Nova Science Publishers, Inc.; 2015.

Hell, M.; Bernhofer, C.; Huhulescu, S.; Indra, A.; Allerberger, F.; Maass, M.; Hakalehto, E. (2010). How safe is colonoscope-reprocessing regarding *Clostridium difficile* spores? *The Journal of Hospital Infection* Vol. 76, Supplement 1: Abstracts, 8th International Congress of the Hospital Infection Society, 10-13 October 2010, Liverpool, UK, S21-22.

Heitto, L.; Heitto, A.; Hakalehto, E. (2009). Tracing wastewaters with faecal enterococci. Poster presentation in 2nd European Large Lakes Symposium, held in Norrtälje, Sweden, 2009.

Laitiomäki, E.; Pesola, J.; Hakalehto, E. (2014). Projected improvement in the fast microbiological analysis of neonatal blood samples, *Journal of Neonatal Nursing*, http://dx.doi.org/10.1016/j.jnn.2014.10.005.

Mentu, J. V.; Heitto, L.; Keitel, H. V.; Hakalehto, E. (2009). Rapid Microbiological Control of Paper Machines with PMEU Method. *Paperi ja Puu / Paper and Timber*, *91*, 7-8 (90th Anniversary n:o).

Pesola, J.; Hakalehto, E. (2011). Enterobacterial microflora in infancy - a case study with enhanced enrichment. *Indian J Pediatr, 78*, 562-568. http://dx.doi.org/10.1007/s12098-010-0341-5.

Pesola, J.; Vaarala, O.; Heitto, A.; Hakalehto, E. (2009). Use of portable enrichment unit in rapid characterization of infantile intestinal enterobacterial microbiota. *Microb Ecol Health Dis., 21*, 203-210. http://dx.doi.org/10.3109/08910600903367810.

Pesola, J.; Paakkanen, H.; Hakalehto, E. (2012). Enhanced diagnostics of pyelonephritis – a case study. *Int. J Med. Med. Sci* 2: 273-277.

Pitkänen, T.; Bräcker, J.; Miettinen, I. T.; Heitto, A.; Pesola, J.; Hakalehto, E. (2009). Enhanced enrichment and detection of thermotolerant *Campylobacter* species from water using the Portable Microbe Enrichment Unit (PMEU) and real-time PCR, *Canadian Journal of Microbiology* 55(7); 849–858.

Wirtanen, G.; Salo, S. (2010). 41st R3-Nordic Symposium. Cleanroom technology, contamination control and cleaning, Dipoli, Espoo, Finland, May 25 - 26, 2010. VTT, Espoo. 223 p.VTT Symposium : 266.

WHO, (2011). Guidelines for drinking-water quality - 4th ed.

In: Coliforms
Editor: Geoffrey McCoy
ISBN: 978-1-63483-542-8

Chapter 4

PREVALENCE OF *ESCHERICHIA COLI* AND OUTBREAKS FROM READY TO EAT FRESH FRUITS AND VEGETABLES

Agni Hadjilouka, Spiros Paramithiotis and Eleftherios H. Drosinos*

Laboratory of Food Quality Control and Hygiene, Department of Food Science and Human Nutrition, Agricultural University of Athens, Athens, Greece

ABSTRACT

Food poisoning outbreaks linked to disease-causing strains of the bacterium *Escherichia coli* are usually associated to meat products. However, 20-30% of *E. coli* infections are caused by consumption of fresh produce due to mishandling at processing facilities, transportation and food service establishments. *E. coli* has been isolated from a variety of products and its prevalence has been reported to range from 2.5% to 93.3% in leafy greens, spinach, lettuce, cucumbers, rocket, coriander, celery, etc. At the same time various fresh fruits and vegetables have been implicated in outbreaks of *E. coli* infections, with the 1996 Japan outbreak linked to consumption of white (daikon) radish sprouts being the world's largest reported outbreak and the 2011 E.U. outbreak linked to seed sprouts being the outbreak with the highest mortality rate. In this

* Corresponding author: E-mail: ehd@aua.gr; Tel.: +30 210 5294713 Fax: +30 210 5294683.

chapter prevalence of *E. coli* in unprocessed and minimally processed fruits and vegetables as well as *E. coli* outbreaks linked to the consumption of such products are addressed.

1. PREVALENCE OF *ESCHERICHIA COLI* IN FRESH FRUITS AND VEGETABLES

A significant proportion of foodborne pathogens can grow and survive in fresh fruits, vegetables and herbs posing significant hazards in consumers' health. Fresh produce are usually consumed either raw, without any undergoing processes, or after minimal processes, that cause little microbial reduction. Hence, the nature of fresh products' consumption along with their non-sterile growth environment and producers' lack of full control of the conditions in the field, explain the reason why fresh and minimally processed products may be easily contaminated with pathogens (Hadjilouka et al., 2014).

Escherichia coli is a microorganism that can survive in soil for up to 20 months and on roots and crop leaves for long time periods. Furthermore, leaves with higher levels of nitrogen or damaged leaves and fruits support prolonged survival of the microorganism (FAO, 2011). *E. coli* O157:H7, for instance, survived in soil samples for 154 to 217 days in lettuce and parsley fields (Islam et al., 2004) and it was also detected in roots and leaves on lettuce and parsley 77 and 177 days, respectively, after soil's contamination by irrigation water or manure compost (Semenov et al., 2009). Furthermore it was found to survive up to 14 days on shredded lettuce, shredded carrots and sliced cucumbers at 5°C and also to multiply on them at 12-21°C (Lin et al., 1996). Therefore, bad agricultural and manufacturing practices in cultivation, harvest, storage, transport and preparation of the products may result in contamination and fresh produce-associated outbreaks (Tournas, 2005).

Surveys on fresh products have revealed the presence of *E. coli* in various vegetables, fruits and fruits, such as cucumbers, rocket, coriander, celery, alfalfa sprouts, lettuce, watermelon, etc.

Prevalence of *E. coli* has been reported to vary from 2.5% in leafy greens (Khalil et al., 2014) to 93.3% in spinach and lettuce samples (Ozpinar et al., 2013). In general, occurrence of the microorganism varied from 2.5, 7.1% and 11.4% (Abadias et al., 2008) to 86.6%, 85% and 93.3% (Castro-Rosas et al., 2012; Ozpinar et al., 2013).

1.1. Prevalence of *E. coli* in Juices and Raw Fruits

Juices

One hundred unpasteurized beetroot juice samples were collected from public markets in Pachuca, Mexico. Incidence of *E. coli* and diarrheagenic *E. coli* pathotypes (DEPs) was 53 and 9%, respectively. Identified DEPs included enterotoxigenic *E. coli* (ETEC), enteropathogenic *E. coli* (EPEC) and non-O157 Shiga toxin-producing *E. coli* (STEC) (Gómez-Aldapa et al., 2014).

Fruits

Abadias et al., (2008) conducted a survey during 2005-2006 in several retail establishments in the Lleida area in Catalonia, Spain, in order to determine potentially pathogenic bacteria. A total of 300 samples were purchased from 4 supermarkets, including 21 single-ingredient ready-to-eat fruits (apple, pineapple, orange, mango and peach). None of the samples was found positive for *E. coli* and *E.coli* O157:H7 presence.

Viswanathan and Kaur (2001) reported 66.7% (8/24) prevalence in fruits samples collected in Mubai, India. Eight mixed fruit samples, 2 pineapple and 7 watermelon samples were purchased from street vendors and faecal *E. coli* was detected in 3 mixed fruit samples, 2 pineapple and 7 watermelon samples.

1.2. Prevalence of *E. coli* in Raw Vegetables

Celery

Thirty-four samples of celery were collected and tested for *E. coli* O157:H7 presence. Six (18%) of the collected samples were positive of pathogen's presence (FDA, 2013b).

Cucumber

One hundred cucumber samples were collected from supermarkets and street markets in the area of Athens, Greece, and tested for *E. coli* O157:H7 presence. The incidence of *E. coli* O157:H7 was estimated at 3% (Hadjilouka et al., 2015).

Jalapeño and Serrano Peppers

Cerna-Cortes et al., (2012) examined a total of 100 jalapeño and 100 serrano pepper samples collected from 4 large public markets of Pachuca City, Hidalgo State, Mexico. Samples were tested for *E. coli* and diarrheagenic *E.*

coli pathotypes' (DEPs) presence. Prevalence of *E.coli* was finally estimated at 58 and 38% in serrano and jalapeño samples, respectively, while DEPs were identified in 38 (38%) serrano and 14 (14%) and jalapeño samples. Identified DEPs included enterotoxigenic *E. coli* (ETEC) and Shiga toxin-producing *E. coli* (STEC). STEC were isolated from 36% of serrano and 14% of jalapeño samples while ETEC were isolated from 12% of serrano and 2% of jalapeño samples. Presence of both STEC and ETEC was identified in 14 serrano and 2 jalapeño samples, while *E. coli* O157:H7 was not detected in any of the STEC-positive samples.

Leafy Green Vegetables

According to Abadias et al., (2008), *E. coli* was present in 27 (11.4%) of the 236 fresh cut vegetable samples examined in a survey conducted during 2005-2006. Samples consisted of arugula, carrot, corn salad, endive, lettuce and spinach samples. In addition, 28 whole vegetables (5 iceberg, 5 oakleaf, 5 trocadero and 5 romaine lettuces, 3 lettuce hearts and 5 endive samples) were tested for *E. coli*'s presence. *E. coli* was detected in 1 (20%) trocadero sample and 1 (20%) endive samples and its total prevalence for the 28 vegetable samples was estimated at 7.1%. However, none of the isolates was identified as *E. coli* O157:H7, neither in fresh cut vegetables nor in whole vegetable samples.

A total of 486 samples, 378 unpackaged whole conventional and 108 packaged organic leafy greens, were examined for presence of *E. coli.* Samples were purchased from open air local street markets, mobile carts and superstores in the city of Alexandria from April 2013 through April 2014. The studied organic samples included arugula, basil, celery, chives, cilantro, dill, leek, mint, parsley, green/red cabbage, and iceberg lettuce. In addition to these vegetables, the studied conventional samples included radish and spinach *E. coli*. The final prevalence was estimated at 2.5% (28/486) (Khalil et al., 2014).

Lettuce

None of the 151 imported lettuce samples collected from retail market in U.K was positive for the presence of *E. coli* O157:H7 (Little et al., 1999).

Mixed Vegetables

None of the 49 samples of raw and frozen mixed vegetables collected from various chain stores in the U.S were positive for the presence of enteropathogenic *E. coli* (Hall et al., 1967). Contrary to the above results, studies in mixed vegetables indicated the presence of *E. coli.* Eighty nine samples of mixed vegetables were collected and tested in Mexico for *E. coli*'s presence and prevalence was estimated at 19% (Nguyen-the and Carlin, 2000).

Tests conducted from 1991 to 1995 by the Food Laboratory in Florida Department of Agriculture and Consumer Services showed that 11 out of 206 raw field vegetables (5.3%) contained *E. coli*. In addition, 349 samples of prepared raw vegetables were tested for *E. coli* presence and 58 (16.6%) of them were found positive (Lin et al., 1996).

A total of 72 samples of seven types of raw vegetables were obtained from street vendors and tested for various food-borne pathogens in Mumbai, India. The collected samples were mixed vegetables (8), radish (8), cabbage (8), carrot (8), cucumber (8), tomato (8), celery (8), coriander (8) and lettuce (8). *E. coli* was detected in 3 mixed salad samples, 4 radish, lettuce and cabbage samples, 6 carrot samples and 2 coriander samples. The total prevalence was finally estimated at 31.9% (23/72) (Viswanathan and Kaur, 2001).

Prevalence of 16.7% was reported in a survey conducted in retailed fresh produce, in Philippines (Vital et al., 2014). A total of 300 samples were collected from five large local retail air markets and from five supermarkets in order to compare the microbiological quality of the two markets. Sampling was performed for 5 types of vegetable produce that are being consumed uncooked: bell pepper, cabbage, carrot, lettuce and tomato. Fifty samples were collected from each supermarket, 10 of each kind, and 250 vegetable samples were collected from the air markets, 50 of each kind. Prevalence of *E. coli* was 10% in bell pepper, 10% in cabbage, 22% in carrot, 24% in lettuce and 16% in tomato samples purchased from open air markets and 20% in bell pepper, 20% in cabbage, 20% in lettuce and 30% in carrot samples collected from supermarkets. *E. coli* was not detected in tomato samples purchased from supermarkets. Hence, total prevalence of *E. coli* in vegetables from open air market and supermarkets was 16.4% and 18%, respectively.

Ozpinar et al., (2013) analyzed 180 vegetable samples (30 spinaches, 30 lettuces, 30 parsleys, 30 carrots, 30 cucumbers and 30 tomatoes) in order to determine the occurrence of *E. coli* and STEC. The samples were collected randomly from six bazaars in Istanbul from March to December, 2010. Occurrence of *E. coli* was 93.3% in spinach, 93.3% in lettuce, 86.6% in parsley, 43.3% in carrot, 33.3% in cucumber and 13.3% in tomato samples. In addition, out of the 180 samples 13 (7.2%) contained STEC. Serotype O157 was found in parsley, lettuce and carrot; O26 in lettuce, parsley, cucumber and carrot; and O111 and O113 in parsley samples.

Rocket

One hundred rocket samples were collected from supermarkets and street markets in the area of Athens, Greece, and tested for *E. coli* O157:H7 presence. The incidence of *E. coli* O157:H7 was estimated at 7% (Hadjilouka et al., 2015).

Sprouts

Fifteen alfalfa and soybean sprouts were tested for *E. coli* during a survey that was conducted in Lleida in Spain. The incidence of *E. coli* was 40% but none of the isolates was identified as *E. coli* O157:H7 (Abadias et al., 2008).

A total of 24 samples of sprouts were obtained from street vendors and tested for various food-borne pathogens in Mumbai, India. Eight samples of green moong sprouts, 8 samples of brown moong sprouts and 8 samples of channa sprouts were analyzed for microorganism's presence. *E. coli* was detected in 5 green moong samples, 4 brown moong sprouts and 7 channa sprouts the total incidence of *E. coli* was estimated at 66.7% (16/24) (Viswanathan and Kaur, 2001).

1.3. Presence of *E. Coli* in Salads

Mixed Salads

Lin et al., (1996) examined the occurrence of *E. coli* O157:H7 and *E.coli* in 63 mixed salad samples in the U.S. Samples were purchased from 31 food service facilities (4 supermarkets, 14 fast food chain restaurants, and 13 family restaurants). None of them was positive for *E. coli* O157:H7, but *E. coli* was detected in 8 vegetable salads and prevalence was 12.7%. These salads had been purchased from 1 supermarket (2 positive samples), 2 fast food chain restaurants (3 positive samples) and 2 family restaurants (3 positive samples).

From September 2004 to January 2005 a total of 130 ready-to-eat mixed salads were tested for the presence of faecal coliforms and *E. coli* in a survey conducted by Castro-Rosas et al., (2012). Samples were purchased from 2 national chain, 2 local and 2 local small inexpensive restaurants in Pachuca-City, Hidalgo, Mexico. Mixed salads purchased from all the restaurants had similar ingredients: iceberg lettuce, tomato, avocado, onion, cucumber and carrot. Additionally to these ingredients, mixed salads obtained from the local small inexpensive restaurants contained water cress, wheat sprouts and radish. Spinach salads with raw spinach, tomato and mushrooms were also purchased from the national chain restaurants. Of the 130 analyzed samples, 129 (99%)

samples were positive for faecal coliforms and of these in 110 (85%) *E. coli* was detected. Among the 110 *E. coli* positive samples 8 (7%) were contaminated with diarrheagenic *E. coli* pathotypes.

Thirteen ready-to-eat salad samples were found positive for *E. coli* presence in a survey held from February through May, 2010, in Porto region. Fifty samples of bagged ready-to-eat salads belonging to seven brands and available in five national supermarket chains were tested for *E. coli* and prevalence was finally estimated at 26% (13/50) (Campos et al., 2013).

1.4. Presence of *E. Coli* in Raw Herbs

Coriander

Ten samples of coriander were tested for the presence of *E. coli* O157:H7 in Mexico. Two of the collected samples were positive for pathogen's presence and prevalence was 20% (FDA, 2013b).

Prevalence of 20% was estimated in a survey conducted in cilantro samples. Forty-one samples were tested for *E. coli* O157:H7 presence and 8 of them were found positive (FDA, 2013b).

2. *Escherichia coli* Outbreaks Associated with Fruits and Vegetables

Escherichia coli is a bacterium widely distributed in the gastrointestinal tract of humans and warm-blooded animals. It is predominantly facultative anaerobe in the bowel and constitutes part of the natural and essential biota of the physiology of a healthy host. Although most of *E. coli* strains are not pathogenic, they may cause infections in immune compromised hosts, whilst there are also some pathogenic *E. coli* strains that can cause gastrointestinal infections in healthy humans and even death (WHO, 2002).

Based on the mechanism of pathogenicity and the clinical syndromes, pathogenic *E. coli* are specified in six groups: Verotoxigenic or Shiga-toxigenic *E. coli* (VTEC or STEC), enterohaemorrhagic *E. coli* (EHEC), enterotoxigenic *E. coli* (ETEC), enteropathogenic *E. coli* (EPEC), enteroinvasive *E.coli* (EIEC), enteroaggregative *E. coli* (EAEC) and diffusely adherent *E. coli* (DAEC) (CDC, 2014b).

Enterohemorraghic *E.coli* O157:H7 is considered as one of the most important foodborne pathogens. The specific serotype was described as a human pathogen and a cause of food-borne disease in 1982, after causing two outbreaks of hemorrhagic colitis linked to the consumption of hamburgers (Berry and Wells, 2010). Nowadays it has been recognized as an important cause of bacterial diarrhea. Its infectious dose is very low and it can be very dangerous for young children (<5) and elderly people, as its spectrum of illness includes from asymptomatic carriage and non-bloody diarrhea to hemorrhagic colitis, thrombotic thrombocytopenic purpura (TTP), hemolytic uremic syndrome (HUS) and death (Cohen and Gianella, 1992; Harris et al., 2003). Nevertheless, most patients recover from the infection within 10 days (WHO, 2011a). Since 2008, *E. coli* O104:H4 has also been reported to cause severe infections in the EU. The most serious outbreak caused by this serotype occurred in 2011, and was linked to the consumption of sprouted seeds. However, the specific serotype is rarely reported worldwide.

The bacterium can be transmitted to humans mainly through the consumption of contaminated food or water. A big variety of products have been linked to the outbreaks caused by *E. coli,* including undercooked ground meat products, raw milk, cheese, beef and contaminated raw vegetables, sprouts and juices (WHO, 2011a). Cattle appear to be the primary vehicle in transmission to humans at the majority of the outbreaks linked to *E. coli* infections. However, fresh produce can also become easily contaminated with *E. coli* through irrigation water, contaminated seeds, insects, pests, wild and domestic animals and feces that are used as composted manure. Hence, fresh produce have also become an emerging source of *E. coli* infections, causing serious outbreaks that have been linked to the consumption of seed sprouts (Table 1), fresh fruits and vegetables (Table 2), raw salads (Table 3) and unpasteurized juices (Table 4). Human disease caused by *E. coli* has been mainly linked to the consumption of contaminated food and water, but person-to person contact is also an important mode of transmission through the oral-faecal route, among close contacts (families, friends, childcare centers, etc.) (WHO, 2011a).

Table 1. Major reported outbreaks caused by *Escherichia coli*, linked to seed sprouts

Pathogen	Year	Location	Type of sprouts	No. of Cases/deaths	Comments	Reference
E. coli O157:H7	1996	Japan	Radish	6561/2	-	FDA, 2013a
E. coli O157:H7	1997	Japan	Radish	126/0	The pathogen was isolated from leftover sprouts kept in the refrigerator	Taormina et al., 1999
E. coli O157:H7	1997	Michigan & Virginia	Alfalfa	108/0	Sprouts from same seed lot in both states	CDC, 1997a
E. coli O157:NM[a]	1998	California	Clover/Alfalfa	8/0	Sprouts from the same sprouter	Taormina et al., 1999

[a] NM: Non motile

Table 2. Major reported outbreaks caused by *Escherichia coli*, linked to fruits and vegetables

Pathogen	Year	Location/ Place	Type of produce	No. of Cases/deaths	Comments	Reference
E. coli O157:H7	1992	Maine	Vegetables	40/-[a]	-	CSPI, 2000
E. coli O157:H7	1993	Oregon	Cantaloupe	27/0	-	CSPI, 2000
E. coli O157:H7	1993	Oregon / Restaurant	Cantaloupe	9/0	-	Del Rosario and Beuchat, 1995
E. coli O11:H43	1994	Minnesota	Fruits and vegetables	25/-	-	CSPI, 2000
E. coli O157:H7	1995	Montana	Lettuce (leafy green, red, romaine)	>70/-	-	CSPI, 2000
E. coli O157:H7	1995	Idaho	Lettuce (romaine)	21/0	Possibly infected food handler	CSPI, 2000
E. coli O157:H7	1995	Maine/ Scout camp	Lettuce (iceberg)	30/0	Cross contamination with raw hamburger juice	CSPI, 2000

Pathogen	Year	Location/ Place	Type of produce	No. of Cases/deaths	Comments	Reference
E. coli O157:H7	1995	Ontario, Canada/ Acute care hospital	Iceberg lettuce	23/0	Lettuce received in the acute care hospital was heavily spoiled	Preston et al., 1997
E. coli O157:H7	1995	Montana / Retail	Lettuce	70/0	Possible contamination from irrigation runoff or compost used to fertilize the fields. Cattle had access to the stream above the pond used for irrigation.	Ackers et al., 1998
E. coli O157:H7	1996	Connecticut and Illinois	Mesclun lettuce	49/0	The implicated lettuce was traced to a single grower processor. Cattle was found near the lettuce fields.	Hilborn et al., 1999; Tauxe 1997
E. coli O157:H7	1996	Connecticut, Illinois and New York	Lettuce	61/-	-	CSPI, 2000

Table 2. (Continued)

Pathogen	Year	Location/ Place	Type of produce	No. of Cases/deaths	Comments	Reference
E. coli O157:H7	1997	Home	Melon	9/-	-	Raybaudi-Massilia et al., 2009
E. coli O157:H7	1998	Indianapolis/ Restaurant	Coleslaw	33/0	-	FDA, 2013a
E. coli O157:H7	1999	Ohio and Indiana	Lettuce	47/-	-	CSPI, 2000
E. coli O157:H7	1999	Nebraska	Lettuce (iceberg)	72/-	-	CSPI, 2000
E. coli O157:H7	2000	Restaurant	Watermelon	736/-	-	Raybaudi-Massilia et al., 2009
E. coli O157:H7	2000	Grocery store	Red grape	14/-	-	Raybaudi-Massilia et al., 2009
E. coli O157:H7	2001	School	Pear	14/-	-	Raybaudi-Massilia et al., 2009

[a] No data available.

Table 3. Major reported outbreaks caused by *Escherichia coli*, linked to raw salads

Pathogen	Year	Location / Place	Type of salad	No. of Cases/deaths	Comments	Reference
E. coli O6:NM	1993	Rhode Island, New Hampshire	Garden Salad (carrots)	47/0	Possible contamination of carrots used in salads. Carrots came from the same state.	CDC, 1994
E. coli O157:H7	1993	Washington/ Church picnic	Pea salad	16/-[a]	-	Meng et al., 2001
E. coli O157:H7	1995	Alberta, Canada/ restaurant	Caesar salad	37/0	-	FDA, 2013a
E. coli O157:H7	1995	Kansas/ Wedding	Fruit salad/ punch	21/-	-	Meng et al., 2001
E. coli O157:H7	1998	California/ Restaurant	Salad	2/0	-	FDA, 2013a
E. coli O157:H7	1998	Wisconsin	Fruit salad	47 (3 HUS[b])/0	-	FDA, 2013a
E. coli O157:H7	2005	Home	Fruit salad	18/-	-	Raybaudi-Massilia et al., 2009

[a] No data available.

[b] Hemolytic uremic syndrome.

Table 4. Major reported outbreaks caused by *Escherichia coli*, linked to unpasteurized fruit juices

Pathogen	Year	Location/ Place	Type of juice	No. of Cases/deaths	Comments	Reference
E. coli O157:H7	1980	Toronto, Ontario, Canada/ Local market	Apple, unpasteurized	14 (4 HUS[a])/1	Juice purchased from a local market and fair. Juice tasted "bad" or "different."	Steele et al., 1982
E. coli O157:H7	1991	Massachusetts/ Small cider mill	Apple	23 (4 HUS)/0	90% drops used in making juice. Apples were not washed or scrubbed. Cattle raised nearby.	Besser et al., 1993
E. coli (ETEC)	1992	India/ Roadside vendor	Orange	6/0	Two roadside vendors selling fresh squeezed juice, one 6 meters away from the garbage heap.	Singh et al., 1995
E. coli O157:H7	1996	Connecticut / Small cider mill	Apple	14 (3 HUS, 1 HUS &TTP[b])/0	Some drops used in juice. Apples brushed and washed in potable water before juiced using a wooden press. Potassium sorbate (0.1%) added as a preservative.	CDC, 1997b

Pathogen	Year	Location/ Place	Type of juice	No. of Cases/deaths	Comments	Reference
E. coli O157:H7	1996	British Columbia, Canada, California, Colorado, and Washington/ Retail market	Apple	70 (14 HUS)/1	Phosphoric acid based solutions used for wash, may had been used incorrectly or at low concentrations. Possibly poor quality apples, some dropped apples used, apple orchard near cattle/deer.	CDC, 1996; Cody et al., 1999
E. coli O157:H7	1996	British Columbia, Canada, California, Colorado, and Washington/ Retail market	Apple	70 (14 HUS)/1	Phosphoric acid based solutions used for wash, may had been used incorrectly or at low concentrations. Possibly poor quality apples, some dropped apples used, apple orchard near cattle/deer.	CDC, 1996; Cody et al., 1999
E. coli O157:H7	1996	Multiple	Apple	56/-[c]	-	Raybaudi-Massilia et al., 2009
E. coli O157:H7	1997	Farm	Apple	6/-	-	Raybaudi-Massilia et al., 2009

Table 4. (Continued)

Pathogen	Year	Location/ Place	Type of juice	No. of Cases/deaths	Comments	Reference
E. coli O157:H7	1998	Ontario, Canada/ Farm -Home	Apple	14/0	Cattle kept in orchard prior to apple harvest. Water supply on farm not potable. Apples used without further inspection, brushing or washing.	Tamblyn et al., 1999
E. coli O157:H7	1999	Oklahoma	Apple	7/0	Drop apples used. Possible contamination from wild and domestic animal manure.	FDA, 2013a
E. coli O157:H7	1999	Home	Apple	5/-	-	Raybaudi-Massilia et al., 2009

[a] Hemolytic uremic syndrome.
[b] Thrombotic thrombocytopenic purpura.
[c] No data available.

2.1. Major Outbreaks of *Escherichia coli* in the United States and Canada

Multistate Outbreak of E. coli O157:H7 Infections Linked to Fresh Spinach, 2006

A multistate outbreak connected to *E. coli* O157:H7 was reported the autumn of 2006. The outbreak was caused after consumption of fresh spinach and as of 6th October, 2006, 199 people from 26 states were infected with the outbreak strain. One hundred forty-one of the patients (71%) were female and 22 (11%) were children under 5 years old, while 102 of the patients (51%) were hospitalized. Thirty-one (16%) developed hemolytic-uremic syndrome (HUS) and 29% of them were children (<18 years old), 8% were 18 to 59 years old, and 14% were older than 60 years old. Based on the patients with available clinical information, 80% of them became ill between 19th August and 5th September, 2006, with the most people (31%) becoming ill from 30th August to 1st September. Three deaths were connected to the outbreak. A 2-year-old child from Idaho, who died after having developed HUS, and two elderly women, one from Wisconsin and one from Nebraska. The outbreak strain was isolated from 13 spinach packages supplied by patients living in 10 different states. Based on the lot codes of the packages, 11 of them were connected to the same production day of the same manufacturing facility. The rest two packages did not have available lot codes but had the same brand name with the other packages (CDC, 2006).

Multistate Outbreak of E. coli O145 Infections Linked to Shredded Romaine Lettuce, 2010

On 16th April, 2010, public health authorities recognized an outbreak of bloody diarrhea in Michigan caused by *E. coli* O145. The outbreak was subsequently linked to other cases in Ohio, New York, Pennsylvania and Tennessee. Investigation was conducted by public health and agriculture officials of the above states, along with Center of Disease Control and Prevention (CDC) and U.S Food and Drug Administration (FDA). Case-control studies in Michigan and Ohio found significant connections between illness and consumption of romaine lettuce. In addition, lettuce was processed at the same processing facility with lettuce consumed by patients in New York, Pennsylvania and Tennessee. The illnesses were finally linked to a common supplier of romaine lettuce on 27th April, 2010, and as of 28th April, 2010, FDA had determined that the implicated romaine lettuce production lots were produced in late March. The implicated lots were no longer in commerce so no recall was necessary at that time, but laboratory tests on processor's products showed contamination on the latter production days, therefore preemptive recalls were triggered. A recall was also issued by a company that obtained lettuce from the same farm with the processing company linked to the outbreak. The recalled lettuce lots were never implicated in illness.

From 1st March, 2010, until 20th May, 2010, a total of 26 confirmed and 7 probable cases from Michigan (11 confirmed and 2 probable), New York (5 confirmed and 2 probable), Ohio (8 confirmed and 3 probable), Pennsylvania (1 confirmed) and Tennessee (1 confirmed) were linked to this outbreak. Among the 30 patients with available clinical information, 12 (40%) were hospitalized. Three of them developed HUS and no deaths were reported (CDC, 2010).

Multistate Outbreak of E. coli O157:H7 Infections Linked to Romaine Lettuce, 2011

From October 2011 to March 2012, CDC in collaboration with state, local, and federal public health agencies in several states and as well as with the U.S. FDA investigated a multistate outbreak of *E. coli* O157:H7 linked to romaine lettuce. The collaborative investigations indicated that romaine lettuce, sold from several stores of a single grocery store chain (Chain A), was the likely source of illnesses in the outbreak. Ill people reported having purchased salads from grocery stores' salad bars (Chain A) during the time period from 5th to 24th October, 2011. A total of 9 grocery stores of the Chain A was identified by patients that reported having purchased salads a week before becoming ill. Furthermore, based on the available dates of salad purchase, it was indicated that ill people purchased the contaminated salads from the same stores within 4 days. Chain A cooperated with the investigation and voluntarily removed the suspected food items from the salad bars on 26th October, 2011. Romaine lettuce served on salad bars of the grocery stores (Chain A) was originated from a single lettuce processing facility and was distributed to the stores through a single distributor. This indicated that the contamination of the romaine lettuce probably occurred before the product reached the grocery stores (Chain A). The U.S. FDA and several state agencies conducted trace back investigations in order to identify the source of contamination. Trace-back investigations focused on patients who had eaten at salad bars on the grocery stores (Chain A) and patients from university campuses in Minnesota (1 patient) and Missouri (2 patients). Trace-back analysis determined that a single common lot of romaine lettuce that had been harvested in Farm A had been used to supply the grocery stores (Chain A) as well as the university campus in Minnesota during the time of the outbreak. This lot was also provided to a distributor that supplied the university campus in Missouri, but records were not sufficient in order to determine if this lot was finally sent to this university campus. The source of contamination was not indentified at the end, due to the fact that Farm A was no longer in production at the time of the investigation.

As of 21st March, 2012, 58 people from 9 states were infected with the outbreak strain of *E. coli* O157:H7. The number of ill people identified in each state was: Arizona (1), Arkansas (2), Illinois (9), Indiana (2), Kansas (2), Kentucky (1), Minnesota (2), Missouri (38), and Nebraska (1). However, two cases were removed from the case count because based on advanced molecular testing they were not related to this outbreak strain. From the 49 people for whom clinical information was available, illnesses ranged from 9th October, 2011 to 7th November, 2011. Thirty-three (67%) of them were hospitalized and 3 developed HUS. Patient's age ranged from 1 to 94 years old, with a median age of 28 years old. Fifty-nine percent was female and no deaths were reported (CDC, 2011).

Multistate Outbreak of E. coli O26 Infections Linked to Raw Clover Sprouts, 2012

In 2012 public health authorities investigated a multistate outbreak caused by *E. coli* O26, a "non-O157 STEC" *E. coli*. Raw clover sprouts served at Jimmy John's restaurant locations were held responsible for this outbreak. FDA and states conducted trace-back investigations that identified a common lot of clover seeds used to grow the sprouts served at the involved restaurant locations. The implicated sprouts were finally traced in two separate sprouting facilities. Both of them had used the same lot of seed to grow clover sprouts and then had distributed Jimmy John's restaurant locations with this product. On 10th February, 2012, the seed supplier notified the sprouting facilities that had received the contaminated lot of clover seed to stop using it.

As of 3rd April, 2012, 29 people from 11 states were reported to be infected with this outbreak strain of STEC O26. Analytically, the number of people reported in each state was: Alabama (1), Arkansas (1), Iowa (5), Kansas (2), Michigan (10), Missouri (3), Ohio (3), Pennsylvania (1), Washington (1), Wisconsin (1), and West Virginia (1). From the 27 patients with available information, 23 (85%) reported having consumed sprouts at Jimmy John's restaurant locations within 7 days before their illness, while illness onset dates ranged from 25th December, 2011, to 3rd March, 2012. Patients' age ranged from 9 to 57 years old, with a median age of 26 years old.

Seven (24%) of them were hospitalized and 89% of the ill people were female. None of them developed HUS and no deaths were reported (CDC, 2012a).

Multistate Outbreak of E. coli O157:H7 Infections Linked to Organic Spinach and Spring Mix Blend, 2012

An outbreak of *E. coli* O157:H7 that was connected with the consumption of organic spinach and spring mix blend was reported in Connecticut, Massachusetts, New York, Pennsylvania, and Virginia in 2012. Collaborative investigations of state, local, federal public health, agriculture and regulatory agencies linked this outbreak to pre-packaged leafy greens, produced by State Garden of Chelsea, Massachusetts. Thirty-one patients were interviewed in order to obtain information regarding the foods consumed the week before their illness. Thirty (97%) of them reported having eaten pre-packaged leafy greens and 15 (48%) reported having eaten Wegmans brand Organic Spinach and Spring Mix blend the week before their illness. Furthermore, 25 (81%) reported having consumed a variety of different brands of pre-packaged leafy greens produced by State Garden. New York Department Health Wadsworth Center Laboratories isolated the outbreak strain of STEC O157:H7 from 4 leftover packages of Wegmans brand Organic Spinach and Spring Mix blend, collected from 4 patients' homes. State Garden was identified as the common producer; however, the source of contamination was not identified. On 2nd November, 2012, a voluntarily recall was announced for Organic Spinach and Spring Mix blend packages, produced by State Garden. In addition, State Garden issued a consumer advisory regarding the products packaged on 12th and 13th October, 2012.

A total of 33 people were infected with the outbreak strain: 2 patients from Connecticut, 3 from Massachusetts, 26 from New York, 1 from Pennsylvania, and 1 from Virginia. Based on patients' available information illness onset dates of ranged from 18th October to 12th November, 2012. Patients' age ranged from 4 to 66 years old, with a median age of 24 years old. Sixty-three percent was female. Thirteen people were hospitalized and two of them developed HUS, but no deaths were reported (CDC, 2012b).

Multistate Outbreak of E. coli O157:H7 Infections Linked to Ready-to-Eat Salads, 2012

In 2013, Centers of Disease Control and Prevention (CDC) in collaboration with public health officials in several states, the U.S. Food and Drug Administration (FDA) and the U.S. Food Safety and Inspection Service (FSIS) investigated a multistate outbreak of *E. coli* O157:H7 infections.

Epidemiologic and trace back investigations indicated that the outbreak occurred due to consumption of two types of ready-to-eat salads produced by Glass Onion Catering and sold at Trader Joe's grocery stores. Nineteen (86%) of the 22 interviewed patients reported having shopped at different Trader Joe's grocery stores and 12 people reported having consumed "Field Fresh Chopped Salad with Grilled Chicken" and "Mexicali Salad with Chili Lime Chicken" purchased from Trader Joe's.

On 20th November, 2013, Glass Onion Catering voluntarily recalled about 181,620 pounds of suspected for contamination with *E. coli* O157:H7 ready-to-eat salads and sandwich wrap products with fully cooked chicken and ham. These products, regulated by FSIS, had been produced between 23rd September and 6th November, 2013, and sent to distribution centers that would allocate the products in retail markets in Arizona, California, Nevada, New Mexico, Oregon, Texas, Utah and Washington. An additional recall was announced for ready-to-eat salads and wraps, regulated by FDA, with expiration dates from 23rd September through 14th November, 2013.

Thirty-three people from 4 states were infected with the *E. coli* O157:H7 outbreak strain. Specifically, 1, 28, 1 and 3 patients were identified in Arizona, California, Texas and Washington, respectively. Twenty-two patients were interviewed and based on their information, illness onset dates ranged between 13th October and 5th November, 2013. Seven of them (32%) were hospitalized. Patients' age ranged from 2 to 78 years old, with a median age of 29 years old and 60% of the patients were female. Two people developed HUS and no deaths were reported (CDC, 2013).

Multistate Outbreak of E. coli O121 Infections Linked to Raw Clover Sprouts, 2014

In 2014, a multistate outbreak was connected to *E. coli* O121 (STEC O121). Officials in local, state and federal public health in collaboration with agriculture and regulatory agencies conducted epidemiologic and trace back investigations that indicated that raw clover sprouts produced by Evergreen Fresh Sprouts, LLC of Idaho were the probable source of the outbreak. Thirteen of the 16 interviewed ill people reported having eaten raw clover sprouts during the week before their illness. Based on Washington State Department of Health and the Idaho Department of Health and Welfare, patients reported having consumed sprouts in sandwiches at several restaurants, including Jimmy John's Gourmet Sandwich the Pita Pit and Daanen's Deli restaurants. In addition, FDA's investigation determined that Evergreen Fresh Sprouts, the time period before the outbreak, had supplied sprouts to 7 restaurants where 9 ill people reported eating before their illness. Eight of the people recalled having consumed sprouts.FDA conducted an inspection on Evergreen Fresh Sprouts' facilities on 22nd and 23rd May, on 27th and 30th May and on 6th June, 2014. A lot of insanitary conditions were observed and reported during these inspections. Condensate and irrigation water dripping from rusty valves, rusty and corroded watering system in the mung bean room, tennis rackets with scratches, chips, and frayed plastic, pitchfork with corroded metal and a squeegee with visible corroded metal and non-treated wood were some of the recorded things. As of 1st July, 2014, Evergreen Fresh Sprouts informed FDA that the company had received a new clover seed lot for sprouting purposes and did not have any more the seed lot associated with the outbreak. The company had been previously advised by FDA and CDC to stop using it in a meeting held on 26th June, 2014.

The outbreak caused 19 infections totally, in 6 states: California (1), Idaho (3), Michigan (1), Montana (2), Utah (1), and Washington (11). Among the ill people who provided the date of the beginning of their illness, the onset date ranged from 1st to 20th May, 2014. Patient's age ranged from 11 to 52 years old. The median age of patients was 27 years old and 68% were female. Seven patients were hospitalized but no one developed HUS and no deaths were reported (CDC, 2014a).

2.2. Major Outbreaks of *Escherichia coli* in Europe

Outbreak of E. coliO104:H4 Infections Linked to Sprouted Seeds, 2011

The largest and worst known STEC-associated outbreak in Europe occurred in 2011. *E. coli* O104:H4 was indicated as the causative agent of the outbreak and consumption of fenugreek sprouted seeds were identified as the most likely source of the infection (Frank et al., 2014).

Based on the scientific report of EFSA (2011), on 21st May, 2011, Germany reported an ongoing outbreak of Shiga-toxin producing *E. coli* serotype O104:H4. As of 27th July, 2011, 3165 cases of diarrheal disease caused by *E. coli* O104:H4 were reported, including 17 deaths. Additionally, 904 cases of HUS were reported, including 33 deaths (Table 5). The Robert Koch Institute (RKI) conducted immediately case-control studies which demonstrated that the clinical disease was linked to the consumption of fresh salad vegetables. Later, the German Federal States performed risk-based inspections taking samples from foods, environment, suppliers, sale markets and retail markets and proved an association between the outbreak and the consumption of sprouted seeds. As of 26th July, 2011, none of the food samples were found positive for the outbreak strain.

For the identification of the food source and in order to stop the outbreak, German Task Force EHEC (Enterohaemorrhagic *Escherichia coli*) was established by the German Federal Ministry of Food, Agriculture and Consumer Protection. The German Task Force EHEC consisted of experts from Federal States in Germany, from the Federal Office of Consumer Protection and Food Safety, the Federal Institute for Risk Assessment, the RKI and EFSA scientists.

The epidemiological investigations conducted by the German Task Force EHEC provided strong evidence that the sprouted seeds from a horticultural farm in Lower Saxony (Establishment A) were the vehicle of the infection of the German STEC O104:H4 outbreak. Furthermore, the investigations revealed that the disease occurrence was linked to the consumption of either 2 different sprouted seed mixtures, the "Keimspross-Mischung/ MildeMischung" (germ sprout or mild blend) or the "Würz-Mischung" (spicy blend), that contained 4 and 3 different kinds of sprouted seeds, respectively. Lentil and fenugreek sprouts were the only common ingredients in both mixtures, but lentil sprouts were also used in other sprouted seed mixtures whilst fenugreek sprouts were used almost exclusively for these 2 mixtures and only small amounts of them were sold unmixed.

The sprouting procedure of fenugreek seeds was minutely examined. Based on the known dates of exposures of the cases and the delivery dates of the sprouted seeds to the places of exposure (restaurants, hotels and canteens) the probable production dates were assessed. The assumption that the fresh sprouted seeds were consumed after a maximum of 14 days after their production along with the total epidemiological curve of the outbreak determined that the lot nos. 48088 and 8266 of fenugreek seeds- produced from April to mid-May, 2011, were responsible for the outbreak. Both lots were traced back to a single importer, who imported the seeds from Egypt to Germany.

Three of the employees of the Establishment A developed diarrheal symptoms, and the onset date of their disease was on 6th May, 2011. Due to the fact that the onset of the disease of the first staff member occurred earlier than the consumers' illness onset, it was questioned whether personnel were the origin of the infection. However, the early exposure of the employees was partly justified by the fact that personnel were given a weekly package of sprouted seeds for private use. Nevertheless, the possibility that the employees might have amplified the spread of the infection in the Establishment or that an unidentified carrier was responsible for the spread, were not excluded.

On 24th June, 2011, France reported bloody diarrhea occurrences after an event in the Commune of Bègles near Bordeaux held on 8th June, 2011. As of 26th July, 2011, 15 cases of bloody diarrhea were reported, 11 of the infected people had participated in the event and *E. coli* O104:H4 was confirmed for the 12 patients. *E. coli* O104:H4 was isolated from patients' stool samples and 9 of them (60%) developed HUS. Investigation conducted to identify the causative agent of the infection indicated sprouts as the vehicle of transmission. The possibility of person-to-person transmission of *E. coli* O104:H4 was also reported.

The website of the Ministry of Health and local physicians suggested recommendations for hygiene and infection controls to general populations and particularly to the attendees of the open day. Furthermore, on 5th July a letter was sent to the participants of the open day, highlighting the importance of hygiene measures and safe food preparation in order to reduce the risk of transmission (Aldabe et al., 2011).

The occurrence of the outbreak in France did not have a direct connection with the outbreak in Germany; none of the event's food handlers or people who served the dishes had recently travelled to Germany, and none of the guests had recent contact with someone coming back from Germany. However, the infection was caused by the same STEC strain (Gault et al., 2011). Patients' interviews revealed a significant connection between the infection and the consumption of sprouted seeds (fenugreek, rocket and with mustard).

After the outbreak in France, EFSA set up a task force composed of experts from its staff members, the European Commission, the affected EU Member States, the ECDC, the World Health Organization (WHO) and the Food and Agriculture Organization (FAO). The investigation indicated that between the identified German and French clusters there was a connection through a single fenugreek importer from Egypt. The importer had distributed more than one lot of seeds to Germany the last 48 months.

Based on the technical report of EFSA (2011), the lot no. 48088 of the fenugreek seeds was the most likely common link. However, the implication of other lots was not excluded. The exact point of contamination in the food chain was not determined, but took place before product's distribution by the importer, probably due to natural fertilizers or contaminated irrigation water.

The primary cases of the reported infections observed in Germany and in France, as well as the related cases observed outside Germany but with previous recent travel to Germany, were caused by the contaminated sprouting seeds. However, since in Germany there were thousands of people infected it is assumed that secondary cases occurred by food handlers or through person-to-person transmission (Aldabe et al., 2011; Hauri et al., 2011).

On 26th July, 2011, The Robert Koch Institute announced that the outbreak in Germany was officially over. The last onset date of illness was reported on 4th July, 2011, and cases with onset after 4th July were not considered to have epidemiological connection with the initial outbreak. In total, the outbreak affected 14 European countries causing 4,069 infections (904 HUS and 3,165 non-HUS) and 50 deaths (33 from HUS and 17 from non-HUS cases).

Table 5. Probable and confirmed HUS-STEC and non-HUS STEC cases per European Country, 26 July, 2011 (ECDC, 2011; WHO, 2011b)

Countries	HUS STEC cases (deaths)	Non- HUS STEC cases (deaths)
Austria	1 (0)	4 (0)
Czech Republic	0 (0)	1 (0)
Denmark	10 (0)	16 (0)
France	7 (0)	10 (0)
Germany	857 (32)	3078 (17)
Greece	0 (0)	1 (0)
Luxemburg	1 (0)	1 (0)
Netherlands	4 (0)	7 (0)
Norway	0 (0)	1 (0)
Poland	2 (0)	1 (0)
Spain	1 (0)	1 (0)
Sweden	18 (1)	35 (0)
Switzerland	0 (0)	5 (0)
United Kingdom	3 (0)	4 (0)
Total	904 (33)	3165 (17)

Outbreak of Escherichia coli O104:H4 Infections Linked to Watercress, 2013

On 9th September, 2013, the Public Health England (PHE) automated outbreak detection system mentioned an increase in the number of cases Shiga toxin producing *Escherichia coli* (STEC) serotype O157, phage type $(PT)_2$. STEC O157 PT_2 is the fourth most common STEC phage type reported in England with a peak of cases during the summer months.

From 30th August to 19th September, 2013, 19 cases were identified in total; England (14), Wales (4) and Scotland (1). One case was asymptomatic and 18 cases were symptomatic (17 primary and 1 secondary). Primary cases' onset dates ranged from 17th to 29th August, 2013. Eleven (64.7%) of the 17 primary cases were female, aged from 4 to 87 with a median age of 65 years old. Cases were dispersed across United Kingdom. Seven (36.8%) of the total cases were hospitalized and 14 (77.7%) of the 18 symptomatic cases reported bloody diarrhea. No deaths or HUS were reported.

As of 16th September, information was obtained from the 17 primary symptomatic cases: Of the 17 cases, 15 reported watercress consumption, from which 13 patients reported having purchased the watercress from Retailer A. Seven of them stated that it was pre-packaged, washed and ready to eat. On 12th September, the Outbreak Control Team stated that watercress from Retailer A was the vehicle of the outbreak and the Foods Standards Agency advised Retailer A to recall watercress products. On the same day Retailer A recalled six pre-packed bagged salads with watercress.

Investigations conducted by the Food Standards Agency's food chain identified the supplier and watercress farms that provided watercress to Retailer A during August, 2013. During that time, 10 farms in southern England provided Supplier A with watercress who supplied Retailer A with the product. Eventually no connection was proved between the outbreak strain and Retailer A (Launders et al., 2013).

2.3. Major Outbreaks of Escherichia *coli* in Asia

Outbreak of Escherichia coli O157:H7 Infections Linked to Radish Sprouts, 1996

The largest reported outbreak worldwide caused by *E. coli* bacterium occurred in 1996 in Japan, where consumption of white (daikon) radish sprouts was epidemiologically linked to an outbreak caused by *E. coli* O157:H7. From May to August, 1996, approximately 10,000 cases of *E. coli* O157:H7 infections were reported in Japan.

The outbreak started on 13th July, 1996. The infection was associated with at least 14 different clusters and most of the cases occurred in school-age children. More than 6,000 primary school children were infected in a large outbreak occurred in Sakai City, Osaka Prefecture. In addition to these infections, another outbreak also occurred in a factory in Kyoto, approximately 50 km from Sakai City the same time period.

The outbreak occurred in Sakai City, Osaka, affected 6,309 school children and 92 school staff members in 62 municipal primary schools as well as 160 people that were mainly among family members of the infected school children. As of 26th August, 1996, 565 people were hospitalized, 101 people were diagnosed with HUS and two children, a 10-year-old girl and a 12-year-old girl, died. *E. coli* O157:H7 was detected in many patients' stool specimens (WHO, 1996).

On 17th July, 1996, an outbreak of *E. coli* O157:H7 began, following the Sakai City outbreak. A 24-year-old male factory worker developed diarrhea and was hospitalized in a local clinic. One day later, a second worker was also hospitalized to the clinic with diarrhea. Both patients developed bloody diarrhea and HUS. On 21st July, 1996, a third worker died by HUS-associated encephalopathy. Stool samples collected by all three workers indicated *E. coli* O157:H7 as the causative agent of their illnesses.

During their shifts, workers could consume any of the meals served at the factory cafeteria that was operated by an outside company and all three patients had recently eaten meals from there. Finally, 74 of the 3,155 employees of the factory reported gastrointestinal symptoms in July and the only common eating exposure was at the factory cafeteria.

For this reason, epidemiologic investigation was conducted in order to determine the source of the contamination. All 25 food handlers and all ill workers were interviewed about gastric symptoms and also provided stool samples for bacteriologic tests. Forty-two cases were clinically defined and 5 were confirmed through culture techniques. Six workers had only gastric pain, fever or general fatigue, 5 patients developed bloody diarrhea and 21 developed diarrhea outside the period of the outbreak. Forty-five (96%) of the 47 workers that reported gastrointestinal symptoms had eaten at the factory's cafeteria from 8th to 14th July, 1996. The age of the infected workers ranged from 18 to 61 years old, with a median age of 30 years old and thirty-nine (83%) of them were male.

Cafeteria's kitchen facilities were examined by the regional public health center on 18th July, 1996. One female food handler developed diarrhea on 17th July, but *E. coli* O157:H7 was not isolated from her stool samples or from any of the other food handlers.

As of 26th August, 1996, a total of 9,578 cases of *E. coli* O157:H7 infections had been reported in Japan, causing the death of 11 people. The investigations indicated that the infections in Kyoto factory and Sakai City were linked to the consumption of radish sprouts. The radish sprouts served at the cafeteria were supplied by a single distributor who received sprouts from four growers, one of whom supplied also radish sprouts to the elementary schools where infections were reported. It was also indicated that the radish sprouts served at the factory cafeteria and at primary schools were shipped the same day (9th July, 1996) by the grower (Watanabe et al., 1999). Although *E. coli* O157:H7 was not isolated from white radish sprouts neither in elementary schools nor in factory's cafeteria, it was concluded that they were the vehicle of the outbreak infections.

ACKNOWLEDGMENTS

The authors gratefully acknowledge funding from the European Union Seventh Framework Programme (FP7/2007-2013) under grant agreement n. 289719 (Project QUAFETY).

REFERENCES

Abadias, M; Usall, J; Anguera M; Solsona, C; Viñas, I. Microbiological quality of fresh, minimally-processed fruit and vegetables and sprouts from retail establishments. *Int. J. Food Microbiol.*, 2008, 123, 121-129.

Ackers, M; Mahon, BE; Leahly, E; Goode, B; Damrow, T; Hayes, PS; Bibb, WF; Rice, DH; Barrett, TJ; Hutwagner, L; Griffin, PM; Slutsker, L. An outbreak, of *Escherichia coli* O157:H7 infections associated with leaf lettuce consumption. *J. Infect. Dis.*, 1998, 177, 1588-1593.

Aldabe, B; Delmas, Y; Gault, G; Vendrely, B; Llanas, B; Charron, M; Castor, C; Ong, N; Weill, FX; Mariani-Kurkdjian, P; Terrier, F; Desjardin, M; Simões, J; Le Bihan, B; Combe, C; Rolland, P. Household transmission of haemolyticuraemic syndrome associated with *Escherichia coli* O104:H4, south-western France, June 2011. *Euro Surveill.*2011, 16 (31):pii=19934. Available online: http://www.eurosurveillance.org/ ViewArticle.aspx? ArticleId=19934 Article published on 4 August 2011.

Berry, D; Wells, E. *Escherichia coli* O157:H7: Recent advances in research on occurrence, transmission, and control in cattle and the production environment. *Adv. Food Nutr.*, Res. 2010, 60, 67-118.

Besser, RES; Lett, SM; Weber, JT; Doyle, MP; Barrett, TJ; Wells, JG; Griffin, PM. An outbreak of diarrhea and hemolytic uremic syndrome from

Escherichia coli O157-H7 in fresh-pressed apple cider. *J. Am. Med. Assoc.*, 1993, 269, 2217-2220.

Campos, J; Mourão, J; Pestana, N; Peixe, L; Novais, Carla; Antunes, P. Microbiological quality of ready-to-eat salads: An underestimated vehicle of bacteria and clinically relevant antibiotic resistance genes. *Int. J. Food Microbiol.*, 2013, 166, 464-470.

Castro-Rosas, J; Cerna-Cortés, JF; Méndez-Reyes, E; Lopez-Hernandez, D; Gómez-Aldapa, CA; Estrada-Garcia, Teresa. Presence of faecal coliforms, *Escherichia coli* and diarrheagenic *E. coli* pathotypes in ready-to-eat salads, from an area where crops are irrigated with untreated sewage water. *Int. J. Food Microbiol.*, 2012, 156, 176-180.

Centers of Disease Control and Prevention (CDC). Foodborne outbreaks of enterotoxigenic *Escherichia coli*-Rhode Island and New Hampshire, 1993. Morbidity and Mortality Weekly Report (MMWR), 1994, 43(81), 7-8.

Centers of Disease Control and Prevention (CDC). Outbreak of *Escherichia coli* O157:H7 infections associated with drinking unpasteurized commercial apple juice. British Columbia, California, Colorado, and Washington, October 1996. Morbidity and Mortality Weekly Report (MMWR), 1996, 45(44), 975.

Centers of Disease Control and Prevention (CDC). Outbreaks of *Escherichia coli* O157:H7 infection associated with eating alfalfa sprouts- Michigan and Virginia, June-July 1997. Morbidity and Mortality Weekly Report (MMWR), 1997a, 46(32), 741-744.

Centers of Disease Control and Prevention (CDC). Outbreaks of *Escherichia coli* O157:H7 infection and cryptosporidiosis associated with drinking unpasteurized apple cider - Connecticut and New York, October 1996. Morbidity and Mortality Weekly Report (MMWR), 1997b, 46, 4-8.

Centers of Disease Control and Prevention (CDC). Update on multi-state outbreak of *E. coli* O157:H7 infections from fresh spinach. 2006. http://www.cdc.gov/ecoli/2006/september/updates/100606.htm. Page last updated: October 6, 2006.

Centers of Disease Control and Prevention (CDC). Investigation update: multistate outbreak of human *E. coli* O145 infections linked to shredded romaine lettuce from a single processing facility. 2010. http://www.cdc.gov/ecoli/2010/ecoli_o145/index.html Page last updated: September 2, 2011.

Centers of Disease Control and Prevention (CDC). Investigation Update: Multistate outbreak of *E. coli* O157:H7 infections linked to romaine lettuce. 2011. http://www.cdc.gov/ecoli/2011/ecoliO157/romainelettuce/032312/index.html Page last updated: December 7, 2011.

Centers of Disease Control and Prevention (CDC). Multistate outbreak of Shiga toxin-producing *Escherichia coli* O26 infections linked to raw clover sprouts at Jimmy John's restaurants. 2012a. http://www.cdc.gov/ecoli/2012/O26-02-12/index.html Page last updated: April 3, 2012.

Centers of Disease Control and Prevention (CDC). Multistate outbreak of Shiga toxin-producing *Escherichia coli* O157:H7 infections linked to organic spinach and spring mix blend. 2012b. http://www.cdc.gov/ecoli/2012/O157H7-11-12/index.html. Page last updated: December 10, 2012.

Centers of Disease Control and Prevention (CDC). Multistate outbreak of Shiga toxin-producing *Escherichia coli* O157:H7 infections linked to ready-to-eat salads. 2013. http://www.cdc.gov/ecoli/2013/O157H7-11-13/index.html. Page last updated: December 11, 2013.

Centers of Disease Control and Prevention (CDC). Multistate outbreak of Shiga toxin-producing *Escherichia coli* O121 infections linked to raw clover sprouts. 2014a. http://www.cdc.gov/ecoli/2014/O121-05-14/. Page last updated: August 1, 2014.

Centers of Disease Control and Prevention (CDC).*E. coli* (*Escherichia coli*) - General Information.2014b. http://www.cdc.gov/ecoli/general/ Page last updated: December 1, 2014.

Center for Science in the Public Interest (CSPI). Updated and Revised August 2000. Outbreak alert! Closing the gaps in our federal food-safety net.2000.http://www.cspinet.org/reports/outbreak_alert/outbreak.pdf. Accessed: 30 April, 2015.

Cerna-Cortes, J F; Gómez-Aldapa, CA, Rangel-Vargas, E; Torres-Vitela, MR; Villarruel-López, A; Castro-Rosas, J. Presence of some indicator bacteria and diarrheagenic *E. coli* pathotypes on jalapeño and serrano peppers from popular markets in Pachuca City, Mexico. *Food Microbiol.*, 2012, 32, 444-447.

Cody, SH; Glynn, K; Farrar, JA; Cairns, KL; Griffin, PM; Kobayashi, J; Fyfe, M; Hoffman, R; King, AS; Lewis, JH; Swaminathan, B; Bryant, RG; Vugia, DJ. An outbreak of *Escherichia coli* O157:H7 infection from unpasteurized commercial apple juice. *An. Intern. Med.*, 1999, 130, 202-209.

Cohen, MB; Giannella, RA. Hemorrhagic colitis associated with *Escherichia coli* O157:H7. *Adv. Intern. Med.*, 1992, 37, 173-195.

Del Rosario, BA; Beuchat, LR. Survival and growth of enterohemorrhagic *Escherichia coli* O157:H7 in cantaloupe and watermelon. *J. Food Prot.*, 1995, 58(1), 105-107.

European Center for Disease Prevention and Control (ECDC). Shiga toxin-producing *E. coli* (STEC): Update on outbreak in the EU (27 July 2011, 11:00). 2011. Available online:http://ecdc.europa.eu/en/activities/sciadvice/_layouts/forms/Review_DispForm.aspx?List=a3216f4c-f040-4f51-9f77-a96046dbfd72&ID=602#sthash.IyaL9En0.dpuf. Accessed: 11 May, 2015.

European Food Safety Authority (EFSA). Shiga toxin-producing *E. coli* (STEC) O104:H4 2011 outbreaks in Europe: Taking Stock. *EFSA Journal*, 2011, 9(10):2390. [22 pp.] doi:10.2903/j.efsa.2011.2390. Available online: www.efsa.europa.eu/efsajournal.

Food and Agriculture Organization of the United Nations (FAO). EMPRES Transboundary Animal Diseases Bulletin 39. FAO Animal Production and Health Division. 2011. Available online: http://www.fao.org/3/a-i2530e/i2530e03.pdf. Accessed: 13 May 2015.

Food and Drug Administration (FDA). Analysis and Evaluation of Preventive Control Measures for the Control and Reduction/Elimination of Microbial Hazards on Fresh and Fresh-Cut Produce. Chapter IV: Outbreaks Tables. 2013a. http://www.fda.gov/Food/FoodScienceResearch/SafePracticesforFoodProcesses/ucm091270.htm. Page Last Updated: 6 February 2013.

Food and Drug Administration (FDA). Analysis and Evaluation of Preventive Control Measures for the Control and Reduction/Elimination of Microbial Hazards on Fresh and Fresh-Cut Produce. Chapter IV: Incidence Tables. 2013b. http://www.fda.gov/Food/FoodScienceResearch/ SafePracticesfor FoodProcesses/ucm091268.htm. Page Last Updated: 24 April 2013.

Frank, C; Milde-Busch, A; Werber, D. Results of surveillance for infections with Shiga toxin-producing *Escherichia coli* (STEC) of serotype O104:H4 after the large outbreak in Germany, July to December 2011. *Euro Surveill.* 2014, 19(14):pii=20760. Available online: http://www.eurosurveillance.org/ViewArticle.aspx?ArticleId=20760.

Gault, G; Weill, FX; Mariani-Kurkdjian, P; Jourdan-da Silva, N; King, L; Aldabe, B; Charron, M; Ong, N; Castor, C; Macé, M; Bingen, E; Noël, H; Vaillant, V; Bone, A; Vendrely, B; Delmas, Y; Combe, C; Bercion, R; d'Andigné, E; Desjardin, M; de Valk, H; Rolland, P. Outbreak of haemolyticuraemic syndrome and bloody diarrhoea due to *Escherichia coli* O104:H4, south-west France, June 2011. *Euro Surveill.* 2011,

16(26):pii=19905. Available online:http://www.eurosurveillance.org/ViewArticle.aspx?ArticleId=19905.

Gómez-Aldapa, CA; Rangel-Vargas, E; Bautista-De León, H; Castro-Rosas, J. Presence of non-O157 Shiga toxin-producing *Escherichia coli*, enterotoxigenic *E. coli*, enteropathogenic *E. coli* and *Salmonella* in fresh beetroot (*Beta vulgaris L.*) juice from public markets in Mexico. *J. Sci. Food Agric.*, 2014, 94(13), 2705-2711.

Hadjilouka, A; Paramithiotis, S; Drosinos, EH. Prevalence of *Salmonella* spp. and occurrence of salmonellosis from ready to eat fresh fruits and vegetables. In C.B. Hacket (Ed.), *Salmonella*: prevalence, risk factors and treatment options. Nova Publishers. 2014, 79-110.

Hadjilouka, A; Mantzourani, KS; Katsarou, A; Cavaiuolo, M; Ferrante, A; Paramithiotis, S; Mataragas, M; Drosinos, EH. Estimation of *Listeria monocytogenes* and *Escherichia coli* O157:H7 prevalence and levels in naturally contaminated rocket and cucumber samples by deterministic and stochastic approaches. *J. Food Protect.*, 2015, 78(2), 311-322.

Hall, HE; Brown, DF; Lewis, KH. Examination of market foods for coliform organisms. *Appl. Microbiol.*, 1967, 15, 1062-1069.

Harris, LJ; Farber, JN; Beuchat, LR; Parish, ME, Suslow, TV; Garrett, EH; Busta, FF. Outbreaks associated with fresh produce: incidence, g, and survival of pathogens in fresh and fresh-cut produce. Comprehensive reviews in food science and food safety, 2003, vol 2 (Supplement), 78-141.

Hauri, AM; Götsch, U; Strotmann, I; Krahn, J; Bettge-Weller, G; Westbrock, HJ; Bellinger, O; Uphoff, H. Secondary transmissions during the outbreak of Shiga toxin-producing *Escherichia coli* O104 in Hesse, Germany, 2011. Euro Surveill. 2011; 16(31):pii=19937. Available online:http://www.eurosurveillance.org/ViewArticle.aspx?ArticleId=19937.

Hillborn, ED; Mermin, JH; Mshar, PA; Hadler, JL; Voetsch, A; Wojtkunski, C; Swartz, M; Mshar, R; Lambert-Fair, MA; Farrar, JA; Glynn, MK; Slutsker, L. A multistate outbreak of *Escherichia coli* O157:H7 infections associated with consumption of mesclun lettuce. *J. Am. Med. Assoc.*, 1999.159, 1758-64.

Islam, M; Doyle, MP; Phatak, SC; Millner, P; Jiang, X. Persistence of enterohemorrhagic *Escherichia coli* O157:H7 in soil and leaf lettuce and parsley grown in fields treated with contaminated manure composts or irrigation water. *J. Food Protect.*, 2004, 67, 1365-1370.

Khalil, RKS; Gomaa, MAE; Khalil; MIM.Detection of shiga-toxin producing *E. coli* (STEC) in leafy greens sold at local retail markets in Alexandria, Egypt. *Int. J. Food Microbiol.*, 2015, 197, 58-64.

Launders, N; Byrne, L; Adams, N; Glen, K; Jenkins, C; Tubin-Delic, D; Locking, M; Williams, C; Morgan, D. Outbreak of Shiga toxin producing *E. coli* O157 associated with consumption of watercress, United Kingdom, August to September 2013. *Euro Surveill.* 2013, 18(44):pii=20624. Available online: http://www.eurosurveillance.org/ViewArticle.aspx?ArticleId=20624.

Lin, CM; Fernando, SY; Wei, C. Occurrence of *Listeria monocytogenes*, *Salmonella* spp., *Escherichia coli* and *E. coli* O157:H7 in vegetable salads. *Food Control*, 1996, 7 (3), 135-140.

Little, C; Roberts, D; Youngs, E; de Louvois, J. Microbiological quality of retail imported unprepared whole lettuces: a PHLS Food Working Group Study. *J. Food Protect.*, 1999, 62(4), 325-328.

Meng, J; Doyle, MP; Zhao, T; Zhao, S. Enterohemorrhagic *Escherichia coli.* In: Food Microbiology: Fundamentals and Frontiers. ASM, Washington, DC, 2001, 193-213.

Nguyen-the, C; Carlin, F. Fresh and processed vegetables. The microbiological safety and quality of food. Gaithersburg (MD): Aspen. 2000, 620-84.

Ozpinar, H; Turan, B; Tekiner, IH; Tezmen, G; Gokc, I; Akınede, O. Evaluation of pathogenic *Escherichia coli* occurrence in vegetable samples from district bazaars in Istanbul using real-time PCR. *Lett. Appl. Microbiol.*, 2013, 57, 362-367.

Preston, M; Davidson, R; Harris, S; Thususka, J; Goldman, C; Green, K; Low, D; Proctor, P; Johnson, W; Khakhria, R. Hospital outbreak of *Escherichia coli* O157:H7 associated in a rare phage type-Ontario. *Can. Comm. Dis. Rep.*, 1997, 23, 33-37.

Raybaudi-Massilia, MR; Mosqueda-Melgar, J; Soliva-Fortuny, R; Martin-Belloso, O. Control of pathogenic and spoilage microorganisms in fresh-cut fruits and fruit Juices by traditional and alternative natural antimicrobials. *Compr. Rev. Food Sci F*, 2009, 8, 157-180.

Semenov, AV; van Overbeek, L; van Bruggen, AHC. Percolation and survival of *Escherichia coli* O157:H7 and *Salmonella enterica* serovar *Typhimurium* in soil amended with contaminated dairy manure or slurry. *Appl. Environ. Microbiol.*, 2009, 75, 3206-3215.

Singh, BR; Kulshereshtha, SB; Kapoor, KN.An orange juice borne diarrhoeal outbreak due to enterotoxigenic *Escherichia coli*. *J. Food Sci. Tech.* 1995, 32, 504-506.

Steele, BT; Murphy, N; Arbus, GS; Rance, CP. An outbreak of hemolytic uremic syndrome associated with ingestion of fresh apple juice. *J. Pediatrics*, 1982, 101(6), 963-965.

Tamblyn, S; De Grosbois, J; Taylor, D; Stratton, J. An outbreak of *Escherichia coli* O157:H7 infection associated with unpasteurized non-commercial, custom-pressed apple cider, Ontario, 1998. *Can. Commun. Dis. Rep.*, 1999. 25, 113-117.

Taormina, PJ; Beuchat, LR; Slutsker, L. Infections associated with eating seed sprouts: an international concern. *Emerg. Infect. Dis.*, 1999, 62, 626-634.

Tauxe, RV. Emerging foodborne diseases: an evolving public health challenge. *Dairy Food Environ. Sanit.* 1997, 17(12), 788-795.

Tournas, VH. Spoilage of vegetable crops by bacteria and fungi and related health hazards. *Crit. Rev. Microbiol.*, 2005, 31, 33-44; Available online: http://dx.doi.org/10.1080/10408410590886024.

Viswanathan, P; Kaur, R. Prevalence and growth of pathogens on salad vegetables, fruits and sprouts. *Int. J. Hyg. Environ. Health*, 2001, 203, 205-213.

Vital, PG; Dimasuay, KGB; Widmer, KW; Rivera, WL. Microbiological quality of fresh produce from open air markets and supermarkets in the Philippines. The Scientific World Journal. Available online: http://dx.doi.org/10.1155/2014/219534. Accessed: 20 May, 2015.

Watanabe, Y; Ozasa, K; Jonathan, MH; Griffin, MP; Masuda, K; Imashuku, S; Sawada, T. Factory outbreak of *Escherichia coli* O157:H7 infection in Japan. *Emerg. Infect. Dis.*, 1999, 5(3), 424-428.

World Health Organization (WHO). 1996. Enterohaemorrhagic *Escherichia coli* in Japan - Update. 1996. Available online: http://www.who.int/csr/don/1996_08_28/en/. Accessed: 12 May 2015.

World Health Organization (WHO). Feng, P; Weagant, SD; Grant, MA; Burkhardt, W. Enumeration of *Escherichia coli* and the Coliform Bacteria. Bacteriology Analytical Manual. 2002. Chapter 4. Available online: http://www.fda.gov/food/foodscienceresearch/laboratorymethods/ucm064948.htm. Accessed: 17 May 2015.

World Health Organization (WHO). Enterohaemorrhagic *Escherichia coli* (EHEC). Fact sheet N° 125. 2011a. Available online: http://www.who.int/mediacentre/factsheets/fs125/en/. Accessed: 14 May 2015.

World Health Organization (WHO). Outbreaks of *E. coli* O104:H4 infection: update 30. 2011b. Available online: http://www.euro.who.int/en/health-topics/emergencies/international-health regulations/news/news/2011/07/outbreaks-of-e.-coli-o104h4-infection-update-30. Accessed: 11 May 2015.

In: Coliforms
Editor: Geoffrey McCoy
ISBN: 978-1-63483-542-8

Chapter 5

SOURCES AND SURVIVAL OF COLIFORM BACTERIA IN TEMPERATE FRESHWATERS: A REVIEW

E. Pip*
University of Winnipeg, Winnipeg, Manitoba, Canada

ABSTRACT

The sources and survival of coliform bacteria in freshwater are reviewed. Coliforms are widely used as indicators of water quality and sanitary status, and levels are regulated as standards and legislation in many jurisdictions. However both total (TC) and fecal (FC) coliforms may originate from various environmental sources besides human intestinal flora, and many are nonpathogenic. In surface water, TC and FC may survive for extended periods depending on physical and chemical factors such as ultraviolet light, temperature, pH, turbidity and organic matter. Biotic factors such as protozoa, bacteriophages and algae may also affect survival.

The group of organisms known as coliform bacteria are commonly used to gauge quality and potential for health risk of potable and recreational waters, and are stipulated in water quality standards and legislation of many jurisdictions worldwide. This large heterogeneous group has been loosely

* Corresponding author: e.pip@uwinnipeg.ca.

functionally defined as Gram-negative, non-spore-forming, lactose fermenting bacilli within the family Enterobacteriaceae (Paruch and Maehlum, 2012). Additional criteria may include the production of acid, or presence of the enzyme β-D-galactosidase (see Edberg et al., 2000). The history of the problematic taxonomy of this group has been reviewed by Leclerc et al., (2001). As aerobic and facultative anaerobic organisms, many of the taxa included under the designation of 'total coliforms'(TC) may occur in water and soil. Some may originate from environmental rather than fecal sources; these may present little health risk, or function as opportunistic pathogens of low invasive concern (Archibald, 2000).

A more specific grouping, the 'fecal coliforms'(FC), consisting of thermotolerant or thermotrophic taxa, has been widely accepted as a better indicator of fecal contamination, and therefore pathogenic potential (Leclerc et al., 2001). These bacteria can usually grow and ferment lactose at 44°C. The most common FC is *Escherichia coli*, many strains of which inhabit the gastrointestinal tract of warm-blooded animals and humans. However some members of this group, such as *Enterobacter*, *Citrobacter* and *Klebsiella*, may have nonfecal origins as well, rendering ambiguous the interpretation of their presence in water (Archibald, 2000).

Enteric bacteria that enter freshwater may become altered into resistant non-culturable but viable forms (Jamieson et al., 2002), or may exist as variant strains that cannot ferment lactose (Dufour, 1977 *in* Edberg et al., 2000), which may render them difficult to detect with common incubation methods, and may lead to underestimation of potential health risk. Non-culturable forms require detection through more intensive defined substrate techniques (Zhai et al., 1995). Metabolically inactive forms are of concern because they may persist for much longer periods of time than actively growing cells (Edberg et al., 2000). After residence in fresh water, some coliforms can subsequently be better capable of enduring acidic gastric conditions if ingested (Gauthier and Clement, 1994).

Coliform bacteria in water may present additional health concerns because of the frequent incidence of R-factors and antibiotic resistant strains, often without relation to the type of water source or pollution level (Niemi et al., 1983). The presence of antibiotic resistance patterns has been proposed as a method to assist in discrimination between human and different animal sources of fecal pollution (e.g., Carroll et al., 2005), but due to widespread veterinary and agricultural use, the universal presence of antibiotics in the environment, and temporal instability of resistance factors (see Field and Samadpour, 2007), such differences are inconsistent and of limited practical application (Souza et al., 1999). For example Moss and Frost (1984) found significant multiple antibiotic resistance in *E. coli* from domestic dogs and cats from a large sample of isolates in Australia.

The complex microbial community of the human intestinal tract may include more than 500 bacterial strains (e.g., Wilson, 1997), but *Escherichia coli* is a multivariant and ubiquitous coliform in both human and animal microbiota. It is thus considered to be the best indicator of fecal pollution (Edberg et al., 2000, Paruch and Maehlum, 2012), although it may survive and establish populations in fresh waters (Byappanahalli et al., 2003), and may occur as strains that cannot be genetically matched with specific human or animal sources (McLellan, 2004). Most frequently *E. coli* strains are nonpathogenic, and fulfill beneficial roles in preventing overgrowth of pathogens and aiding in digestion in the gut. However many pathogenic types exist and these are characterized by mode of effecting illness: enteroinvasive, enteroadherent, enteropathogenic, enterotoxigenic, enteroaggregative and enterohemorrhagic. *E. coli* O157 is a particularly serious enterohemorrhagic strain that can result in grave human illness or death (Kudva et al., 1998); it may be spread by infected cattle as well as deer (Coia, 1998). Despite its serious import, O157 strains may be missed or underestimated with some enumeration methods because these bacteria do not grow well above 44°C (Archibald, 2000). While *E. coli* predominate in intestinal flora, other taxa, such as *Enterobacter*, *Citrobacter* and *Klebsiella* are more frequent in sewage and manure (Carroll and Jasper, 1978).

Coliform levels within watersheds are subject to great temporal and spatial variation (Whitlock et al., 2002), presenting challenges for monitoring and assessment. In general, TC counts in surface water increase with duration and amount of rainfall (Stukel et al., 1990, LeChevallier et al., 1991, Pip and Reinisch, 2012). Coliform bacteria that enter surface waters may originate from a large variety of sources, and may be present at elevated concentrations even in waters associated with minimal agricultural, urban or other human activity (Niemi and Niemi, 1991, Fisher et al., 2000).

Wildlife in remote regions may present significant sources. Both warm- and cold-blooded animals harbor coliform bacteria in their guts and feces (Souza et al., 1999). Freshwater fish (Geldreich and Clarke, 1966) and reptiles (Harwood et al., 1999) have been identified among the sources of FC in freshwaters. Waterfowl have been especially studied due to their proximity to surface waters. Gould and Fletcher (1978) reported that daily coliform outputs for some gull species were similar to those of human feces, although salmonellae were not found.

Epilithic periphyton communities may also harbor FC, and some *E. coli* strains may become naturalized constituents of these communities, presenting a reservoir of FC to the overlying water (Ksoll et al., 2007). The latter authors found that a major proportion of *E. coli* strains found in Lake Superior periphyton originated from waterfowl. Coliforms that originate from plants within the aquatic ecosystem are commonly Enterobacteriaceae (*Erwinia*, *Pantoea*). These may invade tissues with pathological outcomes, or exist in more benign endophytic or epiphytic associations (see Leclerc et al., 2001).

Urban sewage outfalls contribute substantial TC and FC to adjacent surface waters (Arnone and Walling, 2007, Ksoll et al., 2007), and counts are significantly higher near urban and industrial areas (Hong et al., 2010), and in streams within sewered compared to nonsewered basins (Young and Thackston, 1999). According to the latter authors, FC levels are related to density of housing as well as of domestic animals. Urban sewage outfalls may contain inadequately treated sewage as well as overflows from combined sanitary and stormwater sewers (Geldreich et al., 1968, Arnone and Walling, 2007). Urban stormwater may be contaminated by domesticated and feral pets and urban wildlife (Whitlock et al., 2002, Ram et al., 2007). Enteric bacteria have been shown to be capable of growth in river water downstream of a sewage outfall (Hendricks, 1972). FC concentrations from both human and nonhuman sources are elevated during and after stormwater runoff (Van Donsel et al., 1967). Conventional activated sewage sludge systems do not effectively eliminate coliforms such as *Salmonella* and *Pseudomonas* (Archibald, 2000), and may harbor high FC concentrations that can remain viable for several months. Although repeated cycles of freezing and thawing promote mortality, some FC may survive the first winter (Edmonds, 1976). Survival in sludge is greater at higher temperatures and pH (Edmonds, 1976). Human waste may also enter adjacent surface waters from municipal sewage lagoons, septic tanks (Crabill et al., 1999, Chabaud et al., 2006) or failed holding tanks. Brandes (1978) found high concentrations of TC and FC in both gray and black water from septic tanks.

Industrial effluents with no fecal inputs may contribute viable coliforms to receiving waters, for example food-processing industry wastes (Hirn, 1980, Menon, 1985). Both TC and FC are typically present in effluent of pulp and paper mills: *Klebsiella*, *Enterobacter*, *Citrobacter* and *E. coli* may live and multiply in such effluent (Archibald, 2000, Gauthier and Archibald, 2001, Beauchamp et al., 2006). Temperatures of 40-60°C. characteristic of many mill water systems may select for thermotolerant strains (Gauthier and Archibald, 2001). According to the latter authors, however, these FC may not necessarily be pathogenic strains and thus may not reflect health risk. Coliforms in tannery effluent may show both antibiotic and metal resistance (Verma et al., 2001).

Agricultural operations constitute a major source of coliform contamination (Jamieson et al., 2003). Irrigation water may contain appreciable levels of TC and FC which may gain access to surface and ground waters and pose health risks (Entry and Farmer, 2001). Surface water that receives runoff from pastures and farmyards (Collins and Rutherford, 2004), or that is directly accessed by cattle contains substantially higher FC concentrations compared to forested areas (Davies-Colley et al., 2004). According to the latter authors, cattle defecate significantly more frequently during stream crossings than on land, and additional bacteria are introduced with washing of soiled submerged body parts. Intensive livestock production facilities generate particularly large amounts of concentrated waste that constitute a substantial source of TC and FC (Entry et al., 2000 Mallin, 2000, Mallin and Cahoon, 2003). Bacterial transport to surface water occurs via water flow as surface runoff or movement through soil macropores (e.g., Howell et al., 1996) or as aerosols, even at recommended manure to soil application rates (Burkholder et al., 2007). Crane et al., (1981) and Zhai et al., (1995) found that different application rates did not affect bacterial survival in surface-applied manure. Similarly a study of spray vs. injection application methods yielded inconclusive results with respect to survival rates (Cook et al., 1997 *in* Jamieson et al., 2002). However numbers of *E. coli* declined as length of composting time of manure increased before its application to soil (Larney et al., 2003).

A characteristic of bacterial temporal patterns in surface waters impacted by livestock waste is the occurrence of spikes of short duration, which are easily missed by environmental compliance monitoring (Mallin, 2000, Pip and Reinisch, 2012). Fisher et al., (2000) found that FC near poultry and cattle production areas were elevated after rainfall. Similarly Pip and Reinisch (2012) reported that TC inputs from cropland spread with hog and poultry manure increased in adjacent streams after precipitation, and observed that ratios of FC relative to TC escape increased with higher rainfall. In general, greater soil moisture and runoff following rainfall was correlated with higher TC and FC (Hagedorn et al., 1978, Entry et al., 2000, Ramos et al., (2006).

The use of vegetation buffer zones to mitigate overland inputs has been reviewed by Norris (1993). However riparian buffer strips offer inconsistent performance in mitigating TC and FC in surface flows (Entry et al., 2000, Fajardo et al., 2001). Thus limiting manure spreading to periods of drier weather is an important determinant in reduction of TC and FC contamination of surface water from agricultural sources.

Recreational use may also present a significant source of bacterial contamination. Sherry (1986) found that FC concentrations in water increased on days when numbers of bathers were high. Conversely Whitman and Nevers (2003) found no correlation between bathers and *E. coli* counts, but Calderon et al., (1991) reported that rates of gastrointestinal illnesses were correlated with numbers of swimmers, although not with coliform counts. FC may be released by bathers accidentally, originate from wash-off (Anderson et al., 1998, Gerba, 2000), or in the case of young children, from diapers. Even infrequent incidents of fecal release can have significant impacts on pathogen concentrations in the water (Gerba, 2000). Pets, particularly dogs, may soil water and beaches (Wright et al., 2009). Since recreational areas, and particularly beaches, often attract gulls and other waterfowl, bird feces may contribute additional coliform loads, although results are inconsistent (Standridge et al., 1979, Whitman and Nevers, 2003, Ishii et al., 2007). Rainfall events may also be associated with higher coliform counts in recreational waters (Tunnicliff and Brickler, 1984).

Solar inactivation is likely the primary factor in the death of FC in surface waters (Davies et al., 1995), although TC are more resistant (Dewedar and Bahgat, 1995). Survival of *E. coli* decreases exponentially with increase in radiation intensity and decrease in wavelength, with 265 nm being the most effective (Vermeulen et al., 2008). Suspended particles have been reported to be associated with higher TC and FC levels (Hong et al., 2000, Pip and Allegro, 2010, Pip and Reinisch, 2012) and longer survival times (Sherer et al., 1992, Edberg et al., 2000). Particles larger than the 1-6 μm typical size of coliform bacteria that are suspended in water may harbor bacteria and protect them from ultraviolet light (Emerick et al., 2000, Cantwell and Hofmann, 2008). Since rainfall and surface runoff events have been shown to be correlated with greater microbial loads (Kistemann et al., 2002), and suspended solids are associated with such events (Pip and Reinisch, 2012), elevated bacterial counts may thus be exacerbated and prolonged.

Sedimentation is an important process that contributes to the attrition of FC in the water column (Gannon et al., 1983). Bottom sediments act as substantial reservoirs for TC and FC, with continuous exchange between upper sediments and the overlying water (Ishii et al., 2007). Fecal bacteria may persist longer in sediments than in water (Whitman and Nevers, 2003), or even increase there (Laliberte and Grimes, 1982). Sediments may achieve concentrations as high as 1000 times more FC than the overlying water (Xu et al., 1982). The persistence of coliform bacteria in sediments may be attributed in part to shielding from the deleterious action of sunlight (Alm et al., 2003), as well as higher nutrient concentrations. Whitman and Nevers (2003) found that at temperate freshwater beaches, *E. coli* concentrations were highest in foreshore sand, and that these bacteria may be able to maintain their populations during summer in this habitat even without additional external inputs. Alm et al., (2003) observed *E. coli* concentrations up to 38 times higher in the top 20 cm of sand in the wave-washed zone than in the water column, and a positive correlation was seen between *E. coli* counts in the top 5 cm of sand and the overlying water. Whitman and Nevers (2003) also observed that *E. coli* concentrations were greater in shallow water sediments compared to those farther offshore. Wave action, longshore currents, seiches, storms, or human activities such as splashing or boating (Crabill et al., 1999, An et al., 2002) as well as associated shoreline erosion may enable resuspension of particle-bound bacteria into the water column. Where livestock have direct access to waterways, they can produce sediment plumes that resuspend FC (Davies-Colley et al., 2004).

However Reed (1997) reported that solar inactivation in water is dependent on the presence of oxygen, which is often reduced in sediments and the sediment-water interface. While some of the attrition is due to cell damage and death, some may also be due to part of the population adopting a nonculturable, but still viable, state (Barcina et al., 1990). The adverse effect of light on survival is enhanced in waters of higher salinity (Davies and Evison, 1991).

A number of coliform bacteria are capable of survival or growth at low temperatures (1-4°C.)(i.e., psychrotrophs), as well as at higher animal and human body temperatures, for example certain members of *Enterobacter*, *Serratia*, *Rahnella and Buttiauxella* (see Leclerc et al., 2001). Enterohemorrhagic *Escherichia coli* O157:H7 may survive for extended periods in cold water, with maximum survival reported at 8°C (Wang and Doyle, 1998). While higher water temperatures reduce coliform survival (Reddy et al., 1981, Flint, 1987), in practice higher TC counts are often observed at warmer water temperatures (Entry et al., 2000, Hong et al., 2000, Pip and Allegro, 2010, Pip and Reinisch, 2012), which may possibly be confounded by greater inputs at these times. Pip and Allegro (2010) reported that FC counts were enhanced relative to TC at lower temperatures.

Survival times may be affected by water chemistry, which may in turn be influenced by rainfall, or flood or spring thaw events. Values of pH that are both higher (>7) (Curtis et al., 1992) and lower (<6) (Ellis and McCalla, 1976 *in* Jamieson et al., 2002) are associated with lower coliform survival. However in natural surface waters TC is often positively correlated with pH (Hong et al., 2000, Pip and Reinisch, 2012); a possible explanation for this observation is that adverse affects of higher pH may be overshadowed by other associated favorable factors such as nutrient availability (e.g., Davies et al., 1995). Pip and Reinisch reported an increase of TC relative to FC at higher pH (Pip and Reinsich, 2012).

Nutrient availability may also affect survival and even regrowth (Gerba et al., 1975, Davies et al., 1995). Organic matter particles and dissolved organic nutrients are often correlated with coliform counts in water (LeChevallier et al., 1991, Hong et al., 2000, Pip and Allegro, 2010, Pip and Reinisch, 2012). Organic matter may originate from autochthonous sources such as exudates and decomposition products of algae, macrophytes and other biota, or from allochthonous inputs (Cole, 1982). While correlations between coliform counts and organic concentrations may have reflected in part their common pollution sources, organic matter has been shown to increase the survival of *E. coli* (Tate, 1978). Organic matter may contribute towards enhanced survival by providing nutrition, as well as adsorption of bacteria onto particles, and absorption of harmful radiation via protective molecules such as humic and fulvic acids (Davies and Evison, 1991).

Inorganic phosphorus has been found to be correlated with FC in surface water (Hong et al., 2000, Pip and Reinisch, 2012). The latter authors also found correlations between nitrate and FC; Zhai et al., (1995) suggested that low nitrogen availability may contribute to greater FC mortality. The effects of inorganic nitrogen and phosphorus, however, are difficult to assess since TC and FC inputs are generally accompanied by nutrient loadings from the same sources (Pip and Reinisch, 2012).

Domek et al., (1984) studied copper, lead and cadmium in distribution water and found that copper may injure coliform populations by affecting, at least in part, the respiratory chain, with experimental Cu concentrations of 25 μg/L causing >90% injury within 6 days. Pip and Allegro (2010) examined these same metals in Lake Winnipeg, but detected no effect on TC or FC at the somewhat lower environmental Cu concentrations (up to 20 μg/L); however *in situ* studies of any long-range effects are easily confounded by other coincident factors.

Bacteriophage attack may further contribute to coliform die-off in natural waters (Verstraete and Voets, 1976, Burton et al., 1987). The presence of bacterivorous protozoan predators in natural surface waters normally results in net attrition of FC (Davies et al., 1995) and their impact on *E. coli* removal can be substantial (Enzinger and Cooper, 1976, Chabaud et al., 2006). However some coliform bacteria may survive predation by protozoa (Barker and Brown, 1994), including *E. coli* 0157 (Barker et al., 1999). Such endogenous bacteria may therefore show increased resistance to inhospitable aquatic environments (King et al., 1988), and bacteria harbored within amoebae have been associated with coliform persistence in chlorine-treated drinking water (Goshko et al., 1983, King et al., 1988). Bacteria adsorbed onto suspended particles may be protected from predation (Davies and Bavor, 2000).

Contaminated surface waters generally harbor both coliform bacteria and algae. The interactions between aquatic bacteria and algal blooms has been reviewed by Doucette (2006). McFeters et al., (1978) observed coliform outbreaks associated with *Chlorella* in otherwise unpolluted streams in Wyoming, and that FC could grow and assimilate organic compounds from *Chlorella*. According to Mezrioui et al., (1994), die-off of *E. coli* was reduced in the presence of cyanobacteria. Whitman et al., (2003) found that *E. coli* could survive more than 6 months in sun-dried mats of *Cladophora* on Lake Michigan beaches, and thus could present an environmental source of FC. Pip and Allegro (2010) found no correlation between microcystin and either TC or FC in Lake Winnipeg. Bacterial metabolites, for example the pigment prodigiosin from *Serratia*, may inhibit the growth of cyanobacteria, and to a lesser extent, eukaryotic green algae (Darveau and Lynch, 1977). During periods of algal blooms, increased turbidity may also contribute to extended coliform survival.

In conclusion, coliform bacteria in surface waters are expected to increase as anthropogenic sources multiply with human population growth and the frequency and intensity of rainfall and storm events associated with climate change rise (Hunter, 2002), but their broad significance as potential health risk sentinels is uncertain. The problem of using coliforms as bacterial indicators is complicated by the diversity of genera included within the description, as well as the great adaptability and large number of genetic strains, ecotypes and biochemical variants within this group, and even within *E. coli* alone. Accordingly their survival in freshwaters, as well as susceptibility to inimical factors, is also highly variable and unpredictable. Since the sources of coliforms are well documented, management strategies to address potential pathogens in surface waters must focus on point-of-source reductions, monitoring and enforcement programs, and public education.

REFERENCES

Alm, E. W., Burke, J. & Spain, A. (2003). Fecal indicator bacteria are abundant in wet sand at freshwater beaches. *Water Res.*, *37*, 3978-3982.

An, Y. J., Kampbell, D. H. & Breidenbach, G. P. (2002). *Escherichia coli* and total coliforms in water and sediments at lake marinas. *Environ. Pollut.*, *120*, 771-778.

Anderson, M. A., Stewart, M., Yates, M. V. & Gerba, C. P. (1998). Modeling the impact of body-contact recreation on pathogen concentrations in a source drinking water reservoir. *Water Res.*, *32*, 3293-3306.

Archibald, F. (2000). The presence of coliform bacteria in Canadian pulp and paper mill water systems – a cause for concern. *Water Qual. Res. J. Canada*, *35*, 1-22.

Arnone, R. D. & Walling, J. P. (2007). Waterborne pathogens in urban watersheds. *J. Water Health*, *5*, 149-162.

Barcina, I., Gonzales, J. M., Iriberri, J. & Egea, L. (1990). Survival strategy of *Escherichia coli* and *Enterococcus faecalis* in illuminated fresh and marine systems. *J. Appl. Bacteriol.*, *68*, 89-198.

Barker, J. & Brown, M. W. R. (1994). Trojan horses of the microbial world: protozoa and the survival of bacterial pathogens in the environment. *Microbiol.*, *140*, 1253-1259.

Barker, J., Humphrey, T. J. & Brown, M. W. R. (1999). Survival of *Escherichia coli* 0157 in a soil protozoan: implications for disease. *FEMS Microbiol. Lett.*, *173*, 291-295.

Beauchamp, C. J., Simao-Beaunoir, A-M., Beaulieu, C. & Chalifour, F-P. (2006). Confirmation of *E. coli* among other thermotolerant coliform bacteria in paper mill effluents, wood chips screening rejects and paper sludges. *Water Res.*, *40*, 2452-2462.

Brandes, M. (1978). Characteristics of effluents from gray and black water septic tanks. *J. Water Pollut. Control Fed.*, *50*, 2547-2559.

Burkholder, J. M., Libra, B., Weyer, P., Heathcote, S., Kolpin, D., Thorne, P. S. & Wichman, M. (2007). Impacts of waste from concentrated animal feeding operations on water quality. *Environ. Health Perspect.*, *115*, 308-312.

Burton, Jr., G. A., Gunnison, D. & Lanza, G. R. (1987). Survival of pathogenic bacteria in various freshwater sediments. *Appl. Environ. Microbiol.*, *53*, 633-638.

Byappanahalli, M., Fowler, M., Shively, D. & Whitman, R. (2003). Ubiquity and persistence of *Escherichia coli* in a midwestern coastal stream. *Appl. Environ. Microbiol.*, *69*, 4549-4555.

Calderon, R. L., Mood, E. W. & Dufour, A. P. (1991). Health effects of swimmers and non-point sources of contaminated water. *Int. J. Environ. Health Res.*, *1*, 21-31.

Cantwell, R. E. & Hofmann, R. (2008). Inactivation of indigenous coliform bacteria in unfiltered surface water by ultraviolet light. *Water Res.*, *42*, 2729-2735.

Carroll, E. J. & Jasper, D. E. (1978). Distribution of Enterobacteriaceae in recycled manure bedding on California dairies. *J. Dairy Sci.*, *61*, 1498-1508.

Carroll, S., Hargreaves, M. & Goonetilleke, A. (2005). Sourcing fecal pollution from onsite wastewater treatment systems in surface waters using antibiotic resistance analysis. *J. Appl. Microbiol.*, *99*, 471-482.

Chabaud, S., Andres, Y., Lakel, A. & Le Cloirec, P. (2006). Bacteria removal in septic effluent: influence of biofilm and protozoa. *Water Res.*, *40*, 3109-3114.

Coia, J. E. (1998). Clinical, microbiological and epidemiological aspects of *Escherichia coli* O157 infection. *FEMS Immunol. Med. Microbiol.*, *20*, 1-9.

Cole, J. J. (1982). Interactions between bacteria and algae in aquatic ecosystems. *Ann. Rev. Ecol.Syst.*, *13*, 291-314.

Collins, R. & Rutherford, K. (2004). Modelling bacterial water quality in streams draining pastoral land. *Water Res.*, *38*, 700-712.

Crabill, C., Donald, R., Snelling, J., Foust, R. & Southam, G. (1999). The impact of sediment fecal coliform reservoirs on seasonal water quality in Oak Creek, Arizona. *Water Res.*, *9*, 2163-2171.

Crane, S. R., Westerman, P. W. & Overcash, M. R. (1981). Die-off of fecal indicator organisms following land application of poultry manure. *J. Environ. Qual.*, *9*, 531-537.

Curtis, T. P., Mara, D. D. & Silva, S. A. (1992). Influence of pH, oxygen, and humic substances on ability of sunlight to damage fecal coliforms in waste stabilization pond water. *Appl. Environ. Microbiol.*, *58*, 1335-1343.

Darveau, R. P. & Lynch, D. L. (1977). The antibiotic activity of prodigiosins against certain blue-green and green algae. *Phycol.*, *16*, 349-351.

Davies, C. M. & Evison, L. M. (1991). Sunlight and the survival of enteric bacteria in natural waters. *J. Appl. Bacteriol.*, *70*, 265-274.

Davies, C. M., Long, J. A. H., Donald, M. & Ashbolt, N. J. (1995). Survival of fecal microorganisms in marine and freshwater sediments. *Appl. Environ. Microbiol.*, *61*, 1888-1896.

Davies, C. M. & Bavor, H. J. (2000). The fate of stormwater-associated bacteria in constructed wetland and water pollution control pond systems. *J. Appl. Microbiol.*, *89*, 349-360.

Davies-Colley, R. J., Nagels, J. W., Smith, R. A., Young, R. G. & Phillips, C. J. (2004). Water quality impact of a dairy cow herd crossing a stream. *N. Zeal. J. Mar. Freshwat. Res.*, *38*, 569-576.

Dewedar, A. & Bahgat, M. (1995). Fate of faecal coliform bacteria in a waste water retention reservoir containing Lemna gibba L. *Water Res.*, *29*, 2598-2600.

Domek, M. J., LeChevallier, M. W. Cameron, S. C. & McFeters, G. A. (1984). Evidence for the role of copper in the injury process of coliform bacteria in drinking water. *Appl. Environ. Microbiol.*, *48*, 289-293.

Doucette, G. J. (2006). Interactions between bacteria and harmful algae: a review. *Nat. Tox.*, *3*, 65-74.

Edberg, S. C., Rice, E. W., Karlin, R. J. & Allen, M. J. (2000). *Escherichia coli*: the best biological drinking water indicator for public health protection. *J. Appl. Microbiol.*, *88*, 106S-116S.

Edmonds, R. L. (1976). Survival of coliform bacteria in sewage sludge applied to a forest clearcut and potential movement into groundwater. *Appl. Environ. Microbiol.*, *32*, 537-546.

Emerick, R. W., Loge, F. J, Ginn, T. & Darby, J. L. (2000). Modeling the inactivation of particle- associated coliform bacteria. *Water Environ. Res.*, *72*, 432-438.

Entry, J. A., Hubbard, R. K., Thies, J. E. & Fuhrmann, J. J. (2000). The influence of vegetation in riparian filterstrips on coliform bacteria: I. Movement and survival in water. *J. Environ. Qual.*, *29*, 1206-1214.

Entry, J. A. & Farmer, N. (2001). Movement of coliform bacteria and nutrients in ground water flowing through basalt and sand aquifers. *J. Environ. Qual.*, 30, 1533-1539.

Enzinger, R. M. & Cooper, R. C. (1976). Role of bacteria and protozoa in the removal of *Escherichia coli* from estuarine waters. *Appl. Environ. Microbiol.*, *31*, 758-763.

Fajardo, J. J., Bauder, J. W. & Cash, S. D. (2001). Managing nitrate and bacteria in runoff from livestock confinement areas with vegetative filter strips. *J. Soil Water Conserv.*, *56*, 185-191.

Field, K. G. & Samadpour, M. (2007). Fecal source tracking, the indicator paradigm, and managing water quality. *Water Res.*, *41*, 3517-3538.

Fisher, D. S., Steiner, J. L., Endale, D. M., Stuedemann, J. A., Schomberg, H. H., Franzluebbers, A. J. & Wilkinson, S. R. (2000). The relationship of land use practices to surface water quality in the Upper Oconee Watershed of Georgia. *Forest Ecol. Manage.*, *128*, 39-48.

Flint, K. P. (1987). The long-term survival of *Escherichia coli* in river water. *J. Appl. Bacteriol.*, *63*, 261-270.

Gannon, J. J., Busse, M. K. & Schillinger, J. E. (1983). Fecal coliform disappearance in a river impoundment. *Water Res.*, *17*, 1595-1601.

Gauthier, F. & Archibald, F. (2001). The ecology of "fecal indicator" bacteria commonly found in pulp and paper mill water systems. *Water Res.*, *35*, 2207-2218.

Gauthier, M. J. & Clement, R. L. (1994). Effect of a short period of starvation in oligotrophic waters on the resistance of enteric bacterial pathogens to gastric pH conditions. *FEMS Microbiol. Ecol.*, *14*, 275-284.

Geldreich, E. E. & Clarke, N. A. (1966). Bacterial pollution indicators in the intestinal tract of freshwater fish. *Appl. Environ. Microbiol.*, *14*, 429-437.

Geldreich, E. E., Best, L. C., Kenner, A. & Van Donsel, D. J. (1968). The bacteriological aspects of stormwater pollution. *J. Water Pollut. Control Fed.*, *40*, 1861-1872.

Gerba, C. P. (2000). Assessment of enteric pathogen shedding by bathers during recreational activity and its impact on water quality. *Quan. Microbiol.*, *2*, 55-68.

Gerba, C. P., Wallis, C. & Melnick, J. L. (1975). Fate of wastewater bacteria and viruses in soil. *J. Irrigat. Drain. Eng.*, *101*, 157-174.

Goshko, M. S., Pipes, W. D. & Christian, R. R. (1983). Coliform occurrence and chlorine residual in small water distribution systems. *J. Am. Water Works Assoc.*, *75*, 309-312.

Gould, D. J. & Fletcher, M. R. (1978). Gull droppings and their effects on water quality. *Water Res.*, *12*, 665-672.

Hagedorn, C., Hansen, D. T. & Simonson, G. H. (1978). Survival and movement of fecal indicator bacteria in soil under conditions of saturated flow. *J. Environ. Qual.*, *7*, 55-59.

Harwood, V. J., Butler, J., Parrish, D. & Wagner, V. (1999). Isolation of fecal coliform bacteria from the Diamondback Terrapin (*Malaclemys terrapin centrata*). *Appl. Environ. Microbiol.* 65, 865-867.

Hendricks, C. W. (1972). Enteric bacterial growth rates in river water. *Appl. Microbiol.*, *24*, 168-174.

Hirn, J. (1980). Indicator bacteria and salmonella in food-processing and domestic effluent. *J.Water Pollut. Control Fed.*, *52*, 48-52.

Hong, H., Qiu, J. & Liang, Y. (2010). Environmental factors influencing the distribution of total and fecal bacteria in six water storage reservoirs in the Pearl River Delta Region, China. *J. Environ. Sci.*, *22*, 663-668.

Hunter, P. R. (2003). Climate change and waterborne and vector-borne disease. *J. Appl. Microbiol. 94*, 37S-46S.

Ishii, S., Hansen, D. L., Hicks, R. E. & Sadowsky, M. J. (2007). Beach sand and sediments are temporal sinks and sources of *Escherichia coli* in Lake Superior. *Environ. Sci. Technol.*, *41*, 2203-2209.

Jamieson, R. C., Gordon, R. J., Sharples, K. E., Stratton, G. W. & Madani, A. (2002). Movement and persistence of fecal bacteria in agricultural soils and subsurface drainage water: a review. *Can. Biosyst. Eng.*, *44*, 1-9.

Janieson, R. C., Gordon, R. J., Tattrie, S. C. & Stratton, G. W. (2003). Sources and persistence of fecal coliform bacteria in a rural watershed. *Water Qual. Res. J. Canada*, *38*, 33-47.

King, C. H., Shotts, Jr., E. B., Wooley, R. E. & Porter, K. G. (1988). Survival of coliforms and bacterial pathogens within protozoa during chlorination. *Appl. Environ. Microbiol.*, *54*, 3023-3033.

Kistemann, T, Classen, T., Koch, C. Dangendorf, F., Fischeder, R, Gebel, J., Vacata, V. & Exner, M. (2002). Microbial load of drinking water reservoir tributaries during extreme rainfall and runoff. *Appl. Environ. Microbiol.*, *68*, 2188-2197.

Ksoll, W. B., Ishii, S., Sadowsky, M. J. & Hicks, R. E. (2007). Presence and sources of fecal coliform bacteria in epilithic periphyton communities of Lake Superior. *Appl. Environ. Microbiol.*, *73*, 3771-3778.

Kudva, I. T., Blanch, K. & Hovde, C. J. (1998). Analysis of *Escherichia coli* 0157:H7 in ovine or bovine manure and manure slurry. *Appl. Environ. Microbiol.*, *64*, 3166-3174.

Laliberte, P. & Grimes, D. J. (1982). Survival of *Escherichia coli* in lake bottom sediment. *Appl. Environ. Microbiol.*, *43*, 623-628.

Larney, F. J., Yanke, L. J., Miller, J. J. & McAllister, T. A. (2003). Fate of coliform bacteria in composted beef cattle feedlot manure. *J. Environ. Qual.*, *32*, 1508-1515.

Lehevallier, M. W., Schulz, W. & Lee, R. G. (1991). Bacterial nutrients in drinking water. *Appl. Environ. Microbiol.*, *57*, 857-862.

Leclerc, H., Mossel, D. A. A., Edberg, S. C. & Struijk, C. B. (2001). Advances in the bacteriology of the coliform group: Their suitability as markers of microbial water safety. *Annu. Rev. Microbiol.*, *55*, 201-234.

Mallin, M. A. (2000). Impacts of industrial-scale swine and poultry production on rivers and estuaries. *Amer. Sci.*, *88*, 26-37.

Mallin, M. A. & Cahoon, L. B. (2003). Industrialized animal production – a major source of nutrient and microbial pollution to aquatic ecosystems. *Pop. Environ.*, *24*, 369-385.

McFeters, G. A., Stuart, S. A. & Olson, S. B. (1978). Growth of heterotrophic bacteria and algal extracellular products in oligotrophic waters. *Appl. Environ. Microbiol.*, *35*, 383-391.

McLellan, S. L. (2004). Genetic diversity of *Escherichia coli* isolated from urban rivers and beach water. *Appl. Environ. Microbiol.*, *70*, 4658-4665.

Menon, A. S. (1985). Salmonellae and pollution indicator bacteria in municipal and food processing effluents and the Cornwallis River. *Can. J. Microbiol.*, *31*, 598-603.

Mezrioui, N., Oudra, B., Oufdou, K., Hassani, L., Loudiki, M. & Darley, J. (1994). Effect of microalgae growing on wastewater batch culture on *Escherichia coli* and *Vibrio cholerae* survival. *Water Sci. Technol.*, *30*, 295-302.

Moss, S. & Frost, A. J. (1984). The resistance to chemotherapeutic agents of *Escherichia coli* from domestic dogs and cats. *Aust. Vet. J.*, *61*, 82-84.

Nicmi, R. M. & Niemi, J. S. (1991). Bacterial pollution of waters in pristine and agricultural lands. *J. Environ. Qual.*, *20*, 620-627.

Niemi, M., Sibakov, M. & Niemela, S. (1983). Antibiotic resistance among different species of fecal coliforms isolated from water samples. *Appl. Environ. Microbiol.*, *45*, 79-83.

Norris, V. (1993). The use of buffer zones to protect water quality: a review. *Water Res. Manage.*, *7*, 257-272.

Paruch, A. M. & Maehlum, T. (2012). Specific features of *Escherichia coli* that distinguish it from coliform and thermotolerant coliform bacteria and define it as the most accurate indicator of faecal contamination in the environment. *Ecol. Indicat.*, *23*, 140-142.

Pip, E. & Allegro, A. (2010). Nearshore fluctuations in water chemistry, microcystins and coliform bacteria during the ice-free season in Lake Winnipeg, Manitoba, Canada. *Ecohydrol. Hydrobiol.*, *10*, 35-43.

Pip, E. & Reinisch, A. (2012). Stream water quality associated with a livestock/poultry production operation in southeastern Manitoba, Canada. *Soil Water Res.*, *7*, 27-35.

Ram, J. L., Thompson, B., Turner, C., Nechvatal, J. M., Sheehan, H. & Bobrin, J. (2007). Identification of pets and raccoons as sources of bacterial contamination of urban storm sewers using a sequence-based bacterial source tracking method. *Water Res.*, *41*, 3605-3614.

Ramos, M. C., Quinton, J. N. & Tyrrel, S. F. (2006). Effects of cattle manure on erosion rates and runoff water pollution by faecal coliforms. *J. Environ. Manage.*, *78*, 97-101.

Reddy, K. R., Khaleel, R. & Overcash, M. R. (1981). Behavior and transport of microbial pathogens and indicator organisms treated with organic wastes. *J. Environ. Qual.*, *10*, 255-266.

Reed, R. H. (1997). Solar inactivation of faecal bacteria in water: the critical role of oxygen. *Lett. Appl. Microbiol.*, *24*, 276-280.

Sherer, B. M., Miner, J. R., Moore, J. A. & Buckhouse, J. C. (1992). Indicator bacterial survival in stream sediments. *J. Environ. Qual.*, *21*, 591-595.

Sherry, J. P. (1986). Temporal distribution of faecal pollution indicators and opportunistic pathogens at a Lake Ontario bathing beach. *J. Great lakes Res.*, *12*, 154-160.

Souza, V., Rocha, M., Valera, A. & Eguiarte, L. E. (1999). Genetic structure of natural populations of *Escherichia coli* in wild hosts on different continents. *Appl. Environ. Microbiol.*, *65*, 3373-3385.

Standridge, J. H., Delfino, J. J., Kleppe, L. B. & Butler, R. (1979). Effect of waterfowl (*Anas platyrhynchos*) on indicator bacteria populations in a recreational lake Madison, Wisconsin. *Appl. Environ. Micrbiol.*, *38*, 547-550.

Stukel, T. A., Greenberg, E. R., Dain, B. J., Reed, F. C. & Jacobs, N. J. (1990). A longitudinal study of rainfall and coliform contamination in small community drinking water supplies. *Environ. Sci. Technol.*, *24*, 571-575.

Tate, R. L. (1978). Cultural and environment factors affecting the longevity of *Escherichia coli* in Histosols. *Appl. Environ. Micrbiol.*, 35, 925-929.

Tunnicliff, B. & Brickler, S. K. (1984). Recreational water quality analyses of the Colorado River corridor in Grand Canyon. *Appl. Environ. Microbiol.* 48: 909-917.

Van Donsel, D. J., Geldreich, E. E. & Clarke, N. A. (1967). Seasonal variations in survival of indicator bacteria in soil and their contribution to storm-water pollution. *Appl. Microbiol.*, *15*, 1362-1370.

Verma, T., Srinath, T., Gadpayle, R. U., Ramteke, P. W., Hans, R. K. & Garg, S. K. 2001. Chromate tolerant bacteria isolated from tannery effluent. *Biores. Technol.*,*78*, 31-35.

Vermeulen, N., Keeler, W. J., Nandakumar, K. & Leung, K. T. (2008). The bactericidal effect of ultraviolet and visible light on *Escherichia coli.* Biotechnol. Bioeng., *99*, 550-556.

Verstraete, W. & Voets, J. P. (1976). Comparative study of *E. coli* survival in two aquatic ecosystems. *Water Res.*, *10*, 129-136.

Wang, G. & Doyle, M. (1998). Survival of enterohemorrhagic *Escherichia coli*)157:H7 in water. *J. Food Protect.*, *6*, 657-775.

Whitlock, J. E., Jones, D. T. & Harwood, V. J. (2002). Identification of the sources of fecal coliforms in an urban watershed using antibiotic resistance analysis. *Water Res.*, *36*, 4273-4282.

Whitman, R. L. & Nevers, M. B. (2003). Foreshore sand as a source of *Escherichia coli* in nearshore water of a Lake Michigan beach. *Appl. Environ. Microbiol.*, *69*, 5555-5562.

Whitman, R. L., Shively, D. A., Pawlik, H., Nevers, M. B. & Bhyappanahalli, M. N. (2003). Occurrence of *Escherichia coli* and enterococci in *Cladophora* (Chlorophyta) in nearshore water and beach sand of Lake Michigan. *Appl. Environ. Microbiol.*, *69*, 4714-4719.

Wilson, K. H. (1997). Biota of the human gastrointestinal tract. In: *Gastrointestinal Microbiology*,Vol. 2. Mackie, R. I., White, R. E. & Isaacson, R. E. ed. Chapman & Hall, New York. 39- 59.

Wright, M. E., Solo-Gabriele, H. M., Elmir, S. & Fleming, L. E. (2009). Microbial load from animal feces at a recreational beach. *Mar. Pollut. Bull.*, *58*, 1649-1656.

Xu, H. S., Robert, N., Singleton, F. L., Attwell, R. W., Grimes, D. J. & Colwell, R. R. (1982). Survival and viability of non-culturable *Escherichia coli* and *Vibrio cholera*e in the estuarine and marine environment. *Microbiol. Ecol.*, *8*, 313-323.

Young, K. D. & Thackston, E. L. (1999). Housing density and bacterial loading in urban streams. *J. Environ. Eng.* *125*, 1177-1180.

Zhai, Q., Coyne, M. S. & Barnhisel, R. I. (1995). Mortality rates of fecal bacteria in subsoil amended with poultry manure. *Biores. Technol.*, *54*, 165-169.

INDEX

β

β-D-galactosidase, viii, 28, 30, 33, 43, 51, 108

A

access, 2, 16, 81, 111, 113
acid, 14, 23, 35, 85, 108
acidic, 108
acute renal failure, 12
adaptability, 115
adhesions, 13
adsorption, 114
adults, 11, 22
aerosols, 111
aesthetics, 16
afebrile, 13
Africa, 23, 49, 60, 65
agar, 31, 33, 37, 49, 50
age, 89, 90, 91, 92, 96, 97
agencies, 88
agglutination, 9
agriculture, 54, 87, 90, 91
AIDS, 3
alfalfa, 72, 76, 99
algae, ix, 107, 114, 115, 117, 118
anaerobe, 77
animal feeding operations, 116
antibiotic, 6, 21, 34, 43, 50, 54, 66, 99, 108, 109, 111, 117, 123
antibiotic resistance, 6, 21, 34, 43, 50, 54, 66, 99, 108, 109, 117, 123
antibody, 10, 41, 43, 47
antigen, 41
antimicrobial therapy, 19
apples, 85, 86
aquifers, 118
Asia, 96
assessment, 18, 33, 35, 37, 45, 109
asymptomatic, 78, 96
Austria, 95
authorities, 3, 38, 87, 89
automate, 63

B

bacterial cells, 9, 36, 42, 60
bacterial enumeration, 28
bacterial pathogens, vii, 2, 4, 10, 18, 31, 39, 45, 116, 119, 120
bacterial strains, 10, 109
bacteriophages, ix, 57, 107
bacterium, ix, 57, 71, 77, 78, 96
Bangladesh, 14
barriers, 5
base, 3, 37
bedding, 117
beef, 46, 78, 120

benefits, 41, 58, 66
benign, 110
bile, 69
biofilms, vii, 2, 3, 5, 6, 7, 8, 10, 14, 15, 16, 17, 18, 19, 20, 21, 23, 24, 25, 31, 36, 47
bioinformatics, 43
biomass, 7
biomolecules, 56
biosensor(s), viii, 28, 31, 39, 40, 41, 43, 44, 45, 46, 48, 49, 50, 51
biotechnology, 54
birds, 28
blood, 11, 12, 13, 70
blood clot, 12
bloodstream, 47
bowel, 77
brain, 12
Brazil, 29, 45
Burkina Faso, 60, 65, 69
by-products, 15, 17

C

cabbage, 74, 75
cadmium, 114
carbon, 16, 41, 45
carbon nanotubes, 16, 41, 45
case study, 70
cattle, 85, 98, 109, 111, 120, 122
CCA, 34
CDC, 19, 77, 79, 83, 84, 85, 87, 88, 89, 90, 91, 92, 99, 100
celery, ix, 71, 72, 73, 74, 75
cell biology, 42
cell surface, 66
central nervous system, 12
challenges, 46, 54, 68, 109
cheese, 78
chemical(s), ix, 2, 3, 6, 15, 17, 34, 35, 40, 41, 42, 56, 59, 61, 107
chemical industry, 2
chemical properties, 35
chemotherapeutic agent, 121
chemotherapy, 3
chicken, 67, 91
childcare, 78
children, 3, 12, 13, 78, 87, 96, 97, 112
China, 120
chitosan, 15
chlorination, 120
chlorine, 4, 6, 15, 19, 21, 24, 25, 63, 115, 119
cholera, 4, 11, 20
chyme, 66
cities, 60
classes, 9
classification, 43
cleaning, 6, 68, 70
climate, 115
climate change, 115
clinical symptoms, 9
clinical syndrome, 77
clusters, 94, 96
coliformic bacteria, vii, 57
coliforms, vii, ix, 2, 7, 10, 14, 16, 20, 34, 36, 45, 50, 57, 60, 61, 65, 76, 99, 107, 108, 110, 115, 116, 117, 120, 121, 122, 123
Coliline PMEU, ix, 54, 58
colitis, 9, 12, 23, 78, 100
collaboration, 88, 90, 91
colonization, 6
commerce, 87
commercial, 36, 99, 100, 104
communities, 2, 5, 6, 14, 16, 19, 110, 120
community, 5, 6, 20, 122
complement, 44
complexity, 16
compliance, 30, 40, 111
complications, 4
composition, 66
compost, 72, 81
composting, 111
compounds, 2, 60, 66
computer, 58, 65
computing, 37, 67
confinement, 118
Congress, 70
constituents, 6, 110
construction, 3, 7

consumers, vii, viii, 1, 4, 15, 53, 72, 93
consumption, vii, ix, 3, 4, 6, 57, 71, 72, 78, 87, 90, 91, 92, 93, 94, 96, 97, 98, 102, 103
containers, 24, 63
contaminated food, 36, 78
contaminated water, 30, 62, 117
contamination, vii, viii, 2, 3, 5, 7, 16, 19, 27, 28, 29, 30, 33, 34, 37, 39, 40, 43, 44, 50, 57, 61, 63, 68, 70, 72, 80, 81, 83, 86, 88, 89, 90, 91, 94, 97, 108, 111, 112, 121, 122
control measures, 14
copper, 41, 114, 118
coriander, ix, 71, 72, 75, 77
correlation(s), 112, 114, 115
corrosion, 16, 20
cost, 36
Council of the European Union, 46
crop(s), 72, 99, 104
cucumbers, ix, 71, 72, 75
cultivation, ix, 39, 53, 54, 55, 56, 64, 65, 72
culture, viii, 10, 17, 22, 27, 28, 30, 31, 33, 35, 36, 37, 38, 39, 43, 47, 54, 61, 66, 68, 97, 121
culture media, 54
culture methods, viii, 27, 28, 31
cycles, 5, 16, 110
cytometry, 42, 46, 48, 51
Czech Republic, 95

D

daikon, ix, 71, 96
dairies, 117
data analysis, 42
data transfer, viii, 53
database, 42
deaths, 79, 80, 81, 82, 83, 84, 85, 86, 87, 88, 89, 90, 91, 92, 95, 96
decay, 6
decomposition, 114
dehydration, 11, 13
Denmark, 95
Department of Agriculture, 75
desorption, 43
detachment, 6, 16
detection, ix, 4, 7, 10, 17, 31, 32, 33, 35, 36, 39, 40, 41, 42, 43, 45, 46, 47, 48, 49, 50, 51, 54, 55, 56, 59, 60, 62, 65, 66, 67, 68, 69, 70, 95, 108
detection system, 40, 42, 95
developed countries, 4, 9, 15
developing countries, vii, 1, 4
Diabetes, 27
diarrhea, 9, 11, 12, 13, 22, 73, 77, 78, 87, 94, 96, 97, 98, 99, 100
diarrheal disease, 5, 9, 92
digestion, 109
discrimination, 108
diseases, vii, 1, 2, 3, 4, 5, 6, 9, 16, 18, 19, 21, 23, 43, 57, 61, 104
disinfection, viii, 2, 4, 5, 6, 7, 8, 15, 16, 17, 19, 21, 24, 25, 58, 60
dispersion, 6
distribution, vii, 1, 2, 3, 5, 6, 7, 10, 14, 15, 18, 20, 21, 22, 23, 24, 47, 58, 61, 91, 94, 114, 119, 120, 122
diversity, 6, 7, 21, 115, 121
DNA, 31, 35, 36, 38, 41, 43, 45, 49
dogs, 109, 112, 121
dosage, 54
draft, 56
drinking water, vii, 1, 2, 4, 7, 14, 15, 17, 18, 19, 20, 21, 22, 23, 24, 25, 34, 39, 45, 47, 50, 115, 116, 118, 120, 122
DW distribution systems, vii, 1, 3
DWDS, vii, 1, 3, 5, 6, 7, 8, 9, 10, 14, 15, 16, 17
dyes, 42

E

E.coli, vii, 61, 63, 65, 73, 74, 76, 77, 78
early warning, 50, 54, 61
ecology, 18, 119
ecosystem, 61, 110
EEA, 28
effluent(s), 110, 116, 117, 120, 121, 122
Egypt, 93, 94, 103

elaboration, 68
electrolysis, 15
electron, 15
electrophoresis, 10
elementary school, 98
ELISA, 40, 41
e-mail, 63
employees, 93, 97
encephalitis, 4
encephalopathy, 97
enforcement, 115
engineering, 39
England, 22, 29, 47, 95, 96
enteritis, 22
Enterococci, viii, 27, 28, 29, 30, 33, 34, 36, 37, 40, 43, 45, 48, 50
enterohemorrhagic *E. coli* O157 H7, 9
environment(s), viii, 2, 7, 14, 16, 17, 20, 21, 22, 49, 50, 51, 54, 55, 58, 60, 63, 72, 92, 98, 108, 115, 116, 121, 122
environment factors, 122
environmental contamination, 45
environmental control, 54
environmental factors, 7
Environmental Protection Agency (EPA), 18, 28, 29, 30, 40, 50, 51
environmental stress, 6
enzyme(s), 10, 30, 31, 33, 35, 38, 41, 108
epidemic, 23, 24
epidemiologic, 91, 97
epidemiological investigations, 93
epithelium, 9
EPS, 5, 6
equipment, 54, 60, 61
erosion, 113, 122
Escherichia coli, viii, ix, 2, 3, 18, 22, 23, 24, 25, 27, 33, 44, 46, 47, 48, 49, 50, 51, 57, 68, 71, 72, 77, 79, 80, 83, 84, 87, 92, 93, 95, 96, 98, 99, 100, 101, 102, 103, 104, 108, 109, 113, 116, 117, 118, 119, 120, 121, 122, 123
eukaryotic, 115
Europe, 19, 51, 92, 101
European Commission, 94
European Parliament, 46
European Union, 28, 29, 46, 98
evidence, 14, 57, 93
excitation, 42
exposure, 6, 28, 31, 44, 45, 57, 93, 97
extraction, 37, 38, 39, 41, 43

F

faecal contamination, viii, 27, 30, 33, 34, 39, 50, 121
false positive, 38
families, 78
family members, 97
farms, 96
fast food, 76
feces, 4, 7, 78, 109, 112, 123
fertilizers, 94
fever, 11, 12, 13, 97
fiber, 46
filters, 48
filtration, 20, 33, 37, 50, 51
fingerprints, 42
Finland, 53, 59, 60, 61, 63, 68, 69, 70
fish, 109, 119
flight, 43
fluctuations, 121
fluorescence, 35, 36, 48, 51, 61, 63
food, ix, 22, 39, 40, 54, 56, 57, 59, 66, 69, 71, 75, 76, 78, 80, 88, 93, 94, 95, 96, 97, 100, 102, 103, 104, 110, 120, 121
Food and Drug Administration (FDA), 73, 77, 79, 82, 83, 86, 87, 88, 89, 90, 91, 92, 101
food chain, 76, 96
food industry, 54
food safety, 102
food service establishments, ix, 71
force, 94
Ford, 19
formation, 3, 5, 7, 8, 14, 21, 41, 58, 66
France, 18, 33, 35, 94, 95, 98, 102
freezing, 110
freshwater(s), vii, ix, 2, 4, 14, 24, 28, 107, 108, 109, 113, 115, 116, 118, 119

fruits, ix, 71, 72, 73, 78, 80, 102, 103, 104
fullerene, 15
funding, 98
funds, 17
fungi, 104

G

garbage, 84
gastroenteritis, 18, 22, 28, 36, 47
gastroenterologist, 44
gastrointestinal tract, 77, 108, 123
gel, 10
gene expression, 23
gene transfer, 6
genes, 6, 7, 23, 99
genetic marker, 39, 47
genome, 36, 37
Georgia, 119
Germany, 92, 93, 94, 95, 101, 102
glucose, 41, 50
glucoside, 33, 50
graph, 64
Greece, 71, 73, 76, 95
green alga, 115, 117
groundwater, 22, 118
grouping, 108
growth, 5, 6, 7, 8, 14, 15, 16, 34, 40, 54, 55, 58, 60, 61, 64, 65, 66, 68, 69, 72, 101, 104, 110, 113, 115, 119
growth modes, 69
growth rate, 7, 119
guidelines, viii, 15, 28, 34, 53

H

habitat, 113
hazards, 72, 104
headache, 11, 13
health, vii, 1, 3, 4, 5, 16, 28, 39, 46, 51, 57, 61, 72, 87, 104, 105, 107, 108, 111, 115
health care, 39
health effects, 46
health problems, vii, 1, 3
health risks, 4, 111
Helicobacter pylori, 4, 22
hemolytic uremic syndrome, 9, 78, 98, 104
hemorrhage, 12
hepatitis, 4
history, 108
HIV, 41
HIV test, 41
homes, 90
host, 6, 9, 11, 47, 77
hotels, 93
housing, 110
human activity, 109
human body, 113
human health, viii, 3, 17, 27, 43, 44
human right, 2
Hunter, 19, 22, 115, 120
hybrid, 17
hybridization, 36
hygiene, 7, 54, 56, 60, 67, 69, 94

I

ideal, viii, 2, 39, 55
identification, 32, 33, 34, 35, 36, 37, 43, 44, 68, 93
image, 21
image analysis, 21
immunocompromised, 3
improvements, 15, 68
in situ hybridization, 49
in vivo, 20
incidence, vii, 1, 3, 5, 73, 76, 102, 108
increased access, 6
India, 34, 50, 73, 75, 76, 84
indicator organisms, viii, 7, 27, 33, 34, 36, 40, 43, 53, 117, 122
industries, 54
industry, 2, 5, 54, 63, 110
infancy, 70
infants, 11
infection, 14, 37, 38, 45, 46, 78, 92, 93, 94, 96, 99, 100, 104, 105, 117
infrastructure, 3
ingestion, 104

ingredients, 76, 93
injure, 114
injury, 114, 118
inoculum, 10
insects, 78
inspections, 92
integrity, 31
interface, 113
international standards, 3
internist, 44
intestinal flora, ix, 107, 109
intestinal tract, 109, 119
intestine, 7
Iowa, 89
iron, 20
irrigation, 60, 65, 72, 78, 81, 92, 94, 102
Islam, 24, 72, 102
isolation, 10, 32, 33
issues, 63
Italy, 18

J

Japan, ix, 9, 71, 79, 96, 97, 104

K

kill, 5
kinetics, 23, 24

L

laboratory tests, 88
lactose, 108
lakes, 14, 28, 122
leafy greens, ix, 71, 72, 74, 90, 103
leakage, 28
LED, 63
legislation, vii, viii, ix, 27, 28, 107
lettuce, ix, 71, 72, 74, 75, 76, 81, 87, 88, 98, 99, 100, 102
light, ix, 42, 107, 112, 113, 117, 122
liquids, 58
Listeria monocytogenes, 102, 103
livestock, 111, 113, 118, 121
living conditions, 54
locus, 10
longevity, 122
longitudinal study, 122
low temperatures, 113
Luxemburg, 95
lysis, 43

M

macropores, 111
magnitude, 33
majority, 3, 60, 78
malaise, 11
MALDI, 43
management, 3, 46, 68, 115
manufacturing, 72, 87
manure, 72, 78, 86, 102, 103, 109, 111, 117, 120, 122, 123
mapping, 43
marine environment, 123
mass, 43, 47
mass spectrometry, 43
materials, 20
matrix, 5, 43
matter, 60, 114
measurement, 21, 35, 67
meat, ix, 40, 67, 71, 78
media, 10, 33, 34, 35, 37, 38, 48
median, 89, 90, 91, 92, 96, 97
medical, 10, 41
medicine, 56
meningitis, 4
messages, 65
metabolism, 55, 66, 68
metabolites, 38, 67, 115
metabolomic, viii, 28
metals, 114
methodology, 69
Mexico, 73, 75, 76, 77, 91, 100, 102
microarray technology, 35
microbial cells, 6, 58, 68
microbial communities, vii, 2, 5
microbial community, 20, 61, 109

microbiota, 70, 109
microorganism(s), vii, 1, 4, 5, 6, 7, 16, 19, 28, 30, 33, 35, 36, 37, 49, 56, 57, 72, 76, 103, 118
microscopy, 36
miniature, 55
Missouri, 88, 89
mobile communication, 55
moisture, 111
molds, 60
molecular methods, 28, 36, 37, 43
molecules, 38, 39, 41, 114
Montana, 80, 81, 92
mortality, ix, 71, 110, 114
mortality rate, ix, 71
mRNA, 36
mucoid, 13
mucosa, 11
mucus, 11, 12
mung bean, 92

N

nanomaterials, 15, 25
nanoparticles, 16, 41, 42, 45
National Academy of Sciences, 20
nausea, 11
nephropathy, 12
Netherlands, 48, 95
New England, 23
nitrification, 5
nitrogen, 72, 114
Norway, 33, 95
nucleic acid, 31, 35, 37, 42
nucleotide sequence, 35
nutrient(s), 6, 7, 10, 16, 31, 56, 113, 114, 118, 120, 121
nutrient concentrations, 113
nutrition, 114

O

obstruction, 6
OECD, 30, 31, 35, 49
officials, 87, 90
Oklahoma, 86
operations, 111
organ, 3
organic compounds, 24, 115
organic matter, ix, 7, 107, 114
organism, 30, 35, 36, 44, 57, 61
oxidation, 40
oxygen, 7, 31, 66, 113, 117, 122

P

pain, 11, 12, 13, 97
participants, 60, 94
pastures, 111
pathogenesis, 21, 60
pathogens, vii, 2, 4, 5, 6, 7, 9, 10, 15, 16, 19, 24, 31, 36, 37, 39, 40, 43, 46, 47, 50, 57, 59, 60, 61, 65, 72, 75, 76, 78, 102, 104, 108, 109, 115, 116, 122
patient care, 56
Pearl River Delta, 120
permit, 6
pests, 78
pH, ix, 7, 41, 62, 107, 110, 113, 117, 119
phage, 95, 103
phenol, 41
phenotype, 7
Philippines, 75, 104
phosphorus, 114
physicians, 94
physiology, 77
plants, 10, 34, 56, 58, 110
platform, 40, 68
Poland, 95
pollutants, 15
pollution, viii, 7, 27, 30, 34, 37, 42, 43, 48, 51, 57, 108, 109, 114, 117, 118, 119, 121, 122
polymerase, 35
polymerase chain reaction (PCR), 10, 22, 35, 37, 38, 47, 65, 70, 103
pools, 60, 66
population, 2, 6, 21, 113, 115
population growth, 2, 115

porous media, 24
Portugal, 1
positive correlation, 113
poultry, 111, 117, 121, 123
precipitation, 111
predation, 115
predators, 6, 114
preparation, 38, 72, 94
preservative, 84
prevention, vii, 1, 3, 68
primary school, 96, 97, 98
prodrome, 12
producers, 72
project, 60, 62
protection, 6, 19, 118
protective role, 14
proteins, 41, 43, 49
proteomic, viii, 28
protozoa, ix, 4, 57, 107, 115, 116, 117, 118, 120
Pseudomonas aeruginosa, 4, 21, 34
public education, 115
public health, vii, 1, 5, 9, 15, 16, 19, 39, 54, 57, 87, 88, 89, 90, 91, 97, 104, 118
public markets, 73, 102
pulp, 110, 116, 119
purification, 39, 61
pyelonephritis, 70

Q

quantification, 10, 31, 36, 39, 41, 42, 43

R

radiation, 15, 112, 114
radish sprouts, ix, 71, 96, 97
rainfall, 34, 57, 109, 111, 112, 113, 115, 120, 122
Raman spectra, 51
Raman spectroscopy, 43
real time, 10, 66
recall, 88, 90, 91, 96
recommendations, 94
recovery, 58, 68, 69
recreation, 46, 116
recreational, vii, 28, 37, 39, 40, 46, 49, 51, 107, 112, 119, 122, 123
recreational areas, 112
recreational water, vii, 28, 37, 39, 40, 49, 51, 107, 112
regrowth, 15, 16, 20, 23, 114
regulations, 105
regulatory agencies, 2, 90, 91
relevance, 5
reliability, 37, 38, 56
renal failure, 9
reprocessing, 70
requirement, 42
RES, 98
researchers, 16
resistance, 4, 6, 8, 14, 21, 108, 111, 115, 119, 121
resources, 4
restaurants, 76, 91, 93, 100
retail, 73, 74, 75, 91, 92, 98, 103
risk(s), vii, 3, 4, 9, 10, 14, 28, 37, 39, 43, 45, 46, 57, 92, 94, 102, 107, 108, 111, 115
risk assessment, 37
risk factors, 102
RNA, 35, 38, 41
rocket, ix, 71, 72, 76, 94, 102
Romania, 14, 24
roots, 72
routes, 57
rubber, 20
runoff, 45, 81, 110, 111, 112, 118, 120, 122

S

safety, 3, 15, 17, 18, 19, 22, 39, 45, 58, 100, 103, 120
salinity, 113
Salmonella, 4, 30, 34, 36, 44, 45, 49, 55, 66, 67, 69, 102, 103, 110, 120
SAP, 17
scarcity, 2
scattering, 43
school, 96, 97, 98

science, 102
scope, 17
security, 54, 56
sediment(s), 7, 15, 34, 112, 113, 116, 117, 118, 120, 122
seed, ix, 71, 78, 79, 89, 92, 93, 104
selectivity, 23, 33, 60
sensing, 39, 41, 44
sensitivity, 4, 35, 36, 37, 38, 39, 41, 43, 44
sensor(s), 40, 41, 60, 64, 67, 69
septic tank, 110, 116
sequencing, 37, 47
Serbia, 48
sewage, 28, 34, 36, 38, 40, 99, 109, 110, 118
shoreline, 113
showing, 32
signalling, 38, 41, 44
silver, 15
sludge, 15, 36, 110, 118
small intestine, 13
solution, 34, 63
South Africa, 29, 49
Spain, 73, 76, 95, 116
speciation, 43
species, 6, 9, 14, 19, 36, 42, 56, 57, 58, 70, 109, 121
spinach, ix, 71, 72, 74, 75, 76, 87, 90, 99, 100
spore, 108
Spring, 90
sprouting, 89, 92, 93, 95
stability, 36, 38
stabilization, 117
staff members, 94, 97
starvation, 119
state(s), 6, 10, 17, 20, 23, 45, 57, 79, 83, 87, 88, 89, 90, 91, 92, 99, 113
sterile, 54, 72
storage, 5, 56, 72, 120
storms, 113
stormwater, 110, 118, 119
stress, 23, 31
stress response, 23
structure, 9, 11, 20, 46, 122
substrate(s), viii, 28, 30, 33, 40, 44, 45, 48, 108
supplier(s), 87, 89, 92, 96
surveillance, 60, 101
survey design, 28
survival, ix, 2, 5, 10, 21, 24, 55, 72, 102, 103, 107, 111, 112, 113, 114, 115, 116, 117, 118, 119, 121, 122, 123
survival rate, 111
susceptibility, 115
Sweden, 20, 34, 70, 95
Switzerland, 95
symbiosis, 61
symptoms, 9, 11, 93, 97
syndrome, 12, 83, 86, 87, 98, 101
synthesis, 42

T

tanks, 110
target, 34, 35, 39, 41, 43
Task Force, 93
taxa, 108, 109
taxonomy, 108
techniques, 15, 16, 20, 31, 35, 36, 37, 44, 47, 48, 97, 108
technologies, viii, 17, 31, 39, 42, 47, 49, 53, 55, 68
technology, viii, 15, 16, 28, 31, 37, 39, 40, 41, 42, 45, 58, 60, 63, 65, 68, 70
temperature, ix, 7, 10, 54, 56, 64, 107
testing, vii, 39, 40, 41, 43, 54, 57, 89
thiosulphate, 63
threats, 16, 54
thrombocytopenia, 12
thrombocytopenic purpura, 12, 78, 86
time periods, 72
titanium, 16
Togo, 51
toxin, 12, 23, 47, 73, 74, 92, 95, 100, 101, 102, 103
traditional practices, 15
transcription, 35
transducer, 39, 46
transmission, 3, 16, 20, 78, 94, 95, 98

transplantation, 3
transport, 55, 59, 72, 111, 122
transportation, ix, 56, 71
treatment, viii, 2, 3, 5, 10, 14, 15, 21, 24, 25, 34, 38, 40, 47, 60, 62, 66, 68, 102, 117
typhoid, 4
typhoid fever, 4

U

ultrasound, 15, 24, 41, 44
uniform, 56
uniform procedures, 56
United Kingdom, 95, 96, 103
United Nations, 101
United States (USA), 9, 10, 18, 19, 20, 23, 28, 44, 61, 63, 68, 69, 70, 87
urban, 45, 109, 110, 116, 121, 122, 123
urbanization, 2
UV radiation, 24

V

validation, 61
vancomycin, 69
variations, 46, 60, 122
vector, 120
vegetables, ix, 60, 65, 71, 72, 74, 75, 78, 80, 92, 98, 102, 103, 104
vegetation, 111, 118
velocity, 6
viruses, 24, 30, 57, 119
visualization, 66
vomiting, 11, 12, 13

W

Wales, 29, 96
Washington, 50, 68, 83, 85, 89, 91, 92, 99, 103
waste, 2, 28, 36, 60, 65, 66, 110, 111, 116, 117, 118
waste water, 2, 28, 36, 60, 65, 66, 117, 118, 119, 121
water chemistry, 113, 121
water flow, viii, 53, 111, 118
water monitoring, ix, 54, 66
water quality, vii, viii, ix, 2, 3, 5, 7, 16, 18, 28, 39, 40, 41, 43, 44, 46, 49, 50, 51, 57, 63, 70, 107, 116, 117, 119, 121, 122
water quality standards, vii, 40, 107
water resources, 49
water supplies, 15
waterborne diseases, vii, viii, 2, 3, 4, 5, 6, 19, 23, 43
waterborne outbreaks, 4, 9, 10, 57
waterborne pathogens, 4, 7, 16, 39
watershed, 120, 123
waterways, 59, 113
wavelengths, 42
well-being, 2
West Africa, 69
wildlife, 110
Wisconsin, 83, 87, 89, 122
wood, 92, 116
workers, 97
World Health Organization (WHO), 3, 4, 15, 18, 19, 28, 29, 51, 57, 70, 77, 78, 94, 95, 97, 104, 105
worldwide, vii, 1, 5, 9, 78, 96, 107